TIME TRIP ON A MOEBIUS STRIP

by

D. Richard Lewis

This is a work of fiction. Names, characters, places, and incidents either are the product of the author's imagination or are used fictitiously. Any resemblance to actual events or locales or persons, living or dead, is entirely coincidental.

ISBN 0-7414-3797-X

Published by:

INFINITY
PUBLISHING.COM

1094 New DeHaven Street, Suite 100
West Conshohocken, PA 19428-2713
Info@buybooksontheweb.com
www.buybooksontheweb.com
Toll-free (877) BUY BOOK
Local Phone (610) 941-9999
Fax (610) 941-9959

Printed in the United States of America

Printed on Recycled Paper

Published August 2007

incredible symmetry, as if a powerful force had compelled me. I say this because this force had in fact been an eerie, yet intoxicating voice calling my name coming from out of the shell's interior, a voice even today I cannot fully explain, and at the time did not dare give a childhood account to my parents knowing they would then make the shell forever off-limits to me.

This shell I entered, I learned that evening, had been a nautilus shell, which some people know is used as a home by a cephalopod mollusk. I thank God to this day the shell had not contained a mollusk of the shell's monstrous size; the slimy creature would have certainly devoured me with its ninety tentacles and nine teeth I knew they possessed.

But you are now probably asking yourself what happened to me after I had ventured into this shell by an overpowering voice summoning me inside its haunting home. Well, that is what I am slowly but surely working myself up to saying. So please bear with me if I take my time in relating this incredible event. Because before I describe to you what happened to me on that most memorable day, and what is still affecting me every second of my life, I must first tell you about Professor Moebius and the mysterious strip that goes by his great-grandfather's name.

The 'Moebius Strip,' as it is called, is a paradox of science even the genius of Albert Einstein failed to fully explain. Nevertheless, it was when I first learned of the strip that I instantly realized the connection the strip had mathematically with the mathematical configuration of the nautilus shell. For both the strip, and this particular species of shell, it seemed, were related in a manner one would never have realized. Because it was only when both were used together in a most unusual way that both of their mysteries would be revealed.

As to what these mysteries are I discovered, I can only tell you this, this early on: that when a Moebius strip is placed within a nautilus shell of the size I foolishly entered as a child, time and space lose all of their secrets, and the meaning of God is suddenly revealed. Yet for this to take place, one has first to ride the strip in a vehicle within this shell, if one is to experience this mind-boggling event. However, that is only half of what would happen to one's reality as a result.

Still, you must be wondering again if I experienced the above when I entered the shell year's later. And what, if anything, happened to me on exiting from the shell's interior having ridden on the strip. Yet for me to answer these questions I must ask you to bear with me once more before I can speak of these things. For I know from experience no one would believe me. Because it is only after you have learned more fully of my involvement with not only my shell, but with the great-grandson of Dr. Moebius, that you will be able to judge the facts: the facts relating to why I had to kill the great-grandson of Dr. Moebius - the only other person on earth able to explain and construct the Moebius strip and its equally mysterious craft.

Chapter Two

As I continue writing in my lab the incredible events of my story, I do so after having sorrowfully killed Professor Henry M. Moebius, who now rests at my feet. The specific reason leading up to why I had to unwillingly kill this most extraordinary man concerns what we had both discovered, as I have previously related.

I first met Professor Moebius while a freshman at college in 1960. He was a teacher of astrophysics then and I was one of his admiring students. It was known throughout the grounds of Harvard, the professor's great-grandfather, whose ancestors had been related to Martin Luther, had been Dr. August Ferdinand Moebius, the famous German mathematician, astronomer, and topologist, who in 1858, had discovered the topological curiosity known as the 'Moebius Strip'. And that Professor Henry M. Moebius' father, Dr. Paul Julius Moebius, had once worked with Dr. Sigmund Freud as a psychiatrist. And as I remember it now, every student in my astrophysics class knew, by way of demonstration by Professor Moebius, that one of the peculiar properties of the Moebius strip, when given a half twist and taped together at its ends, then cut down the middle of its length, resulted in producing, not two separate pieces of paper, as one might have expected, but in only one strip double its original circumference. And then when the same strip was cut and re-taped right afterwards, having been given a full twist, it mysteriously ended up being two interlocking rings of paper when the strip was again cut down the twisted middle of its length.

My interest in this dumbfounding strip of twisted

paper with its mind-boggling enigmas would, to my regret, pursue me for the next few years of my college life and well after. And it was because of this continuing interest with the Moebius strip that I was determined to solve its mysteries before I went frustratingly mad.

It was exactly the same when I went out and bought that equally maddening curiosity - the 'Rubik's Cube'. That crazy checkered cube of colored squares had almost succeeded in giving me a complete nervous breakdown until I had bought the book on how to solve the cube's code. But there was no such book on the mystery of the Moebius strip unfortunately. One was left to one's self on figuring out why the strip behaved like it did. And it was for this reason, among others, that I wanted to become close friends with the great-grandson of Dr. Moebius, in the hope some day, he might aid me to unravel the strip's arcana.

The professor at that time had been a man in his sixties. Yet looking at his dead body at my feet, he now looked to have aged little, incredible as that might seem - his round metal-framed glasses and white bushy mustache only reminding me of my own deceased grandfather.

There seemed to be other resemblances about the professor that reminded me of my grandfather, who had also been a scientist. But whatever they were I was, at the moment, too filled with fear to remember. For I knew I had killed a man and would have to pay for what I had done. But I had first thought of this evil deed some time before, only then I never believed I would have to resort to killing Professor Moebius in order to stop him from doing what I knew all along he might eventually do.

It was in 1980 that Professor Moebius and I had finally succeeded in constructing the Moebius vehicle. And it was about then I knew I might have to kill him should he later do what I feared. The thought of this sin had become fixed in my mind only after I had ridden inside the shell in the vehicle on the two metal tracks of the giant Moebius strip we had assembled.

For it was when I made my first voyage around the mysterious twist of the strip that I began reliving my childhood experience for the second time. It manifested itself in my mind again as if I had ingested some mind-altering drug. But there was also something more that happened to me in the shell. Something once understood by man would most assuredly change the cosmos forever; something even the likes of Dr. Einstein would have not wanted to guess.

Years before I rode the vehicle inside the shell and had this second encounter, Professor Moebius had tried to explain to me certain facts that might clarify his great-grandfather's discovery. He had explained that perhaps, astronomically speaking, space was closed upon itself, and in addition, twisted in the Moebius strip's way. And when I first heard the voice in the shell as a child, I had perhaps been in a state of consciousness where one's individuality had somehow dissolved into an undifferentiated oneness. How my individuality had dissolved into this oneness he was at a loss to explain. But then Professor Moebius had gone on to say, that perhaps the nautilus shell's effect on my mind when a child, had been most likely due to the shell's most perfect symmetry - a symmetry that might well incorporate all the known particles and thus explain the structure of matter.

But what about the unknown eerie voice inside the shell I could not fully describe or explain? The professor then said the voice inside the shell was perhaps being caused by aggregations of atoms that were 'dancing' so to speak. And by their dancing produced sounds when the rhythm of their dancing changed - atoms were known to be harmonic oscillators - their nuclei being the oscillators themselves, the electrons and orbits being seen as the reverberations and echoes of the periodic harmonic motions of the atom's nucleus. And this being the case, it would create dense and subtle forms because of these atomic reverberations - the air inside the cavity of the shell then vibrating very strongly on reaching a certain point its resonating frequency.

I was then reminded of what the Buddhist Yogacara school taught: 'That out of mind spring innumerable things, conditioned by discrimination. These things man accepts as an external world. What appears to be external does not exist in reality; it is indeed mind that is seen as multiplicity; the body, property, and above all these, I say are nothing but mind.'

I little realized then, that telling Professor Moebius what had happened to me while in the shell would only cause the professor to want to experiment with my shell years later - an experiment that might lead to disaster. For I knew someday he would try to duplicate what I had heard in my past, making whatever it had been manifest to his ears as well.

However, little did I suspect the professor would use the Moebius strip, placed in my shell of all places, to accomplish this end. And little did I suspect the dire consequences that would assuredly result. For if the space inside my shell was closed upon itself, as the professor so theorized, what would happen if that space were to suddenly unravel? Would it then cause the space outside the shell to become twisted like the strip, instantly reversing the polarity of living matter? That was something I could never permit, even if the professor had been my own father. But I must admit I was not thinking of stopping the professor in regards to his experiment at that time. I was only thinking of that voice I had heard - a voice that had haunted me for twenty-six years.

Maybe I had not heard the voice those many years because of having hidden the shell within a small cave near the spot it had arrived on shore - this being done before my parents or anyone else had had a chance to discover it.

How I accomplished this extraordinary feat of moving this shell was, at that time quite easy. For knowing all shells of this type were hollow, then discovering the shell had a closed off chamber inside, I had simply let the high tide float the monstrous shell directly into the cave. This I accomplished by guiding the shell with ropes tied to my father's twin-motored speedboat, making sure the shell was afterwards tied down and secured in the cave, so as to prevent the shell from being washed out to sea.

I later learned from reading my many books on science and magnetism that perhaps the shell had washed up near my house because of a terrestrial field of magnetic force located where the fusion between the forces of cosmic radiation had been naturally occurring, which might then have explained the voice I had heard - it resulting from the shell being at this spot and being caused to simply resonate.

Yet that explanation I read was almost exactly what Professor Moebius tried to explain had been the cause of the voice, but in a more complicated manner it then seemed.

However, this theory the professor tried to explain to me had been some fifteen years earlier. And in the ensuing years I kept up a close correspondence in order that we might exchange our ideas about the Moebius strip, as well as the many other related projects we had been doing in which I treasured his advice.

My work, as I previously mentioned, involved doing research in marine biology at a research lab at Port Aransas. I had purposely chosen this particular lab station to do my work for one specific reason: to keep as close as humanly possible to the cave containing my colossal shell. It was very unlikely anyone would ever discover the cave where my shell was residing;

nevertheless, I wanted to be as close to that place as possible - it being the immediate area where my parent's house and childhood home still stood.

For the thirteen years I had been working at my research lab, I had reluctantly postponed revealing my discovery of the shell to anyone other than Professor Moebius. Then suddenly, I had a complete change of mind in the matter, divulging almost on a whim, the shell's mysterious location to my assistants. This I did because of the constant worry I had endured over the years, fearing the shell would eventually be found.

I also told my assistants my long held secret so they would help me bring the shell to the lab. I wanted to achieve some peace of mind and perhaps do some experimentation work on why the shell had produced that haunting voice.

My assistants, however, had a rather difficult time floating my shell out of the cave at high tide, then loading it onto a flatbed truck and bringing it to my lab. Still, the relief of finding a permanent place where my shell would be safe, and where I knew nobody could make a claim to it, was well worth the sacrifice of revealing to them my stupendous find. And they thanked me for at last allowing them to see what they had always thought to be only a wild joke on my part.

Nevertheless, the very moment the shell was safely placed in the back yard of my lab station, I knew I had made a mistake. It was not that I was afraid the shell would be stolen, or that the State of Texas would confiscate the shell for one of its many museums, it was because the shell was suddenly reacting on me in a most uncomfortable way. And it was because of this I immediately made a phone call to Professor Moebius at Harvard University, revealing to him for the first time the location of my long talked about shell.

"Hello, Professor Moebius. This is Dr. Philip Grieg calling. How are you?" I said warmly, feeling at the same time I was talking to my father.

"Oh hello, Philip," I heard him reply. "And how is my number one student and fish doctor?"

"Student? I've been out of a classroom for years, Professor," I said, wondering if he had forgotten me.

"I'm only kidding you, Philip. So, what is it you're calling me about this time? Have you finally discovered why seahorses don't eat hay? Or have you finally figured out the secret of my great-grandfather's strip of twisted paper perhaps?"

"No, unfortunately. I'm calling you about my shell, Professor. It's now at my lab," I told him almost reluctantly. "And I think it's about time you come here and have a look at it."

"You mean you're finally going to show me this 'Eighth Wonder of the World'?'"

"Yes, Professor, I am. And I'm dead serious about it."

"Well it's about time you showed me that mysterious shell of yours. But Philip, how in heaven's name did you manage to move the thing to your lab from wherever it was?"

"With the help of my assistants, Professor, but I think I should have left the shell hidden. Because I've just experienced what I told you about when I first went into the shell as a child."

"But, Philip, are you sure you haven't been inhaling the chloroform you use in those experiments of yours? You know how that stuff can induce hallucinations. Or have you perhaps been reading Walt Whitman's *Leaves of Grass* again, my dear boy?"

"Why, come to think of it, I was reading Whitman's *Leaves of Grass* before I started feeling somewhat strange."

"Well, you know that poem of Whitman's is said to be able to induce sudden revelations of cosmic consciousness in certain individuals. It's even been said Whitman himself was in such a state of consciousness when he wrote that beautiful work. Why, the poem has even been compared to the ancient *Upanishads* and *Bhagavad-Gita* of early India. And Whitman hadn't even read or heard of them until Ralph Waldo Emerson informed him of this fact months after the poem's sale."

"Well Whitman must have been writing his *Leaves of Grass* inside of a nautilus shell, Professor, because I'm beginning to feel sort of cosmic conscious myself. And that's another reason for you to come on down."

"Well in that case, Philip, I accept. Just give me about six hours. I'm just dying to see that shell anyway. Come to think of it, but there might be a connection with your shell's new location and your sudden mental awareness. So just make sure you don't go anywhere near your shell until I first have a chance to examine it. And for heaven's sake, Philip, don't go inside the shell, if you haven't already done so. For God only knows what might happen to you again if you do... Do you understand me, Philip? Don't go into your shell under any circumstances."

"Yes, Professor Moebius, I understand. But please hurry down. Because I'm not sure if I can continue resisting," I said with some dread in my voice.

"What do you mean, Philip?"

"I'm not quite sure, Professor. It's just that I feel like going into the shell all of a sudden, as if that voice were again calling to me to go inside like I was twelve years old."

"Then get out of your lab station's area at once, Philip! You can wait for me at the airport. You must get away from that shell and its influence over you immediately... Do you hear me, Philip?"

"Yes, Professor Moebius," I replied, almost in a daze. "I'll do as you say and leave for the airport now."

After hanging up the phone, I then began making my way to the front door of my lab, wondering to myself why I had called Professor Moebius in the first place. Apart from the fact he was an astrophysicist and mathematician, I did not really know how he was going to help me. What I really seem to have needed at that moment was a psychiatrist. Maybe I had asked the professor to come to the lab because I knew him to be the only person in the world who really knew what had happened to me in my shell. Or maybe it was because I likened the old fellow to my father, who had been a high school history teacher and had died of cancer only three years before - the professor sort of taking the place of both my father and grandfather now that they were dead.

However, as I exited my lab, the awesome sight of my shell began to have a noticeable affect on me. Maybe this was because of the Whitman poem I had been reading in my lab in the morning before calling the professor. What he had told me about that poem's affect on peoples' minds, I already knew from my extensive reading on the subject of mysticism. But it had been Walt Whitman's *Leaves of Grass* that had gotten me interested in things of the occult to begin with.

It had been almost immediately after I had discovered the shell on the beach that I began to read books about self-discovery. And it wasn't until I had read *Leaves of Grass* that I learned the structure of the poem was not primarily logical, but psychological and almost musical in progression, almost as musical as the inside of the shell, which I was to discover later. I say this, because to me, the poem's themes always seemed to change into one another by association, like in a dream - the very state I had seemed to be in when I had first ventured into my strange and enormous crustacean. But as I kept thinking about Whitman's poem, the more I began to wonder what would happen to me if I foolishly disregarded the professor's warning about entering the shell. Would I begin to experience what Sri Ramakrishna had written in his *Gospel* where he had said: 'The Divine Mother revealed to me in the Kali temple that it was She who had become everything. She showed me that everything was full of Consciousness, the Image was Consciousness, the altar was Consciousness, the water-vessels were Consciousness, the doorsill was Consciousness, the marble floor was Consciousness - all was Consciousness.'

Would I too walk out of that temple of a shell another Lord Krishna or another Lord Jesus perhaps? Or would the shell's eerie affect on my mind end up turning me into a walking zombie instead? I was wondering. But there was that mysterious voice I now wanted to hear again, strange as this may sound. Because looking back upon the first day I had heard the voice, I suddenly remembered, as if awakening from a dream, that the voice had actually been a feminine one; a feminine one I would now give almost anything to hear one more frightening time. But I had given my word to Professor Moebius I would not go into the shell under any circumstances until he had had a chance to examine it.

****

Walking into Corpus Christi airport terminal a half hour later, and then waiting for the professor's plane to arrive, I busied myself by reading over to myself again Whitman's *Leaves of Grass*. I had brought this book along with me to determine if it had been his poem or the shell that had caused me to feel strange back at the lab. I had never experienced anything of an unusual nature while reading Whitman's poem before. As I marveled at the futuristic style his poem had been written in, I noticed I was reading it with an interest and mental intensity like never before. There seemed to be passages in the poem that were now beginning to take on more of a profound meaning to me, especially after what the professor had said to me on the phone. And since I already knew about the poem's rapturous affect it seemed to have on certain people, I began searching for particular lines of the poem that might shed some light on the condition I had been in at my lab.

Was it this verse I was now reading? Whitman had written:

Through me forbidden voices,
Voices of sexes and lusts...
Voices veiled, and I remove the veil,
Voices indecent by me clarified and transfigured.

Had it been this verse I might have read only hours before which had made me feel as though I had again heard the voice? I was now trying to decide. Or had I only been undergoing the first telltale symptoms of schizophrenia?

Well there was only one way for me to know that for sure. And that was to re-enter the shell again and find out if I was truly sane, or slowly becoming what I inwardly felt in the deepest interior of my soul: one hell of a mental basket case.

****

Fifty additional minutes passed and I had read the poem twice through without finding another passage like the one I had found. So dropping two quarters into one of those TV chairs, I sat back and watched a quiz show while I waited for the professor's plane to arrive.

On seeing him with his small carryon suitcase in hand approach me some four hours later, I immediately gave him a loving smile and said, "Thank you for coming, Professor. I was beginning to think I shouldn't have called you. I mean, all I really seem to need is a few good sessions with a psychiatrist after what happened this morning."

"Nonsense, dear boy, you're as sane as can be. You only need a drink to settle your nerves. You look awful, like you've seen a ghost."

"Come to think of it, you don't suppose this voice I heard twenty-six years ago was the voice of some ghostly spirit residing in that shell of mine, do you, Professor?" I said, sort of wide-eyed, my mind in more of a worried state than before.

"Well that's sort of why I decided to come and have a firsthand look at this wonder of yours. And like I said to you on the phone, there just might be a correlative relationship between where your shell is now resting and your sudden mental awareness, as far as this mysterious voice is concerned. And if that's the case, Philip, then you may have to wear on your head some kind of a Faraday cage in order to protect you from what I believe to be nothing more than a strong field of terrestrial magnetism, or telluric current coming from the exact spot your shell is now resting, or for that matter, the entire area of Padre Island itself."

"But, Professor Moebius, you've got to be kidding. I can't go around my lab wearing a Faraday cage kind of hat. And besides, what if I actually want to hear that voice again?"

"Philip!" The professor almost shouted. "You can't be serious? I know you've told me many times you wanted to re-experience this childhood encounter of yours, if only to prove to yourself it was a hallucination. But you'd be fooling around with your sanity. For all we know, that voice, or whatever it was, may have been nothing more than a local radio station your mind was strangely receiving. Or God forbid something much worse, that I, being a religious person, must take into account."

"My God! Professor! You can't mean?"

"Yes, Philip. Like maybe a voice coming from Satan himself!"

On hearing Professor Moebius pronounce with quivering lips the name of the devil, I suddenly realized the professor and I might be placing our mortal lives in perhaps the greatest of danger if we decided to go back to my lab and begin experimenting with my shell. But I knew, one way or another, I had to go back there with him and discover, once and for all, the true origin of the shell's mysterious affect on my mind.

Chapter Three

It was on our way to Port Aransas in my '75 Chevy that the professor informed me he had, hours before, phoned Corpus Christi University and asked a certain Dr. Rogers, who he had an acquaintance with, to meet us at my lab.

It so happened that this Dr. Rogers was also a beautiful young woman I would soon fall head over heels in love with. However, little did I know at the time of her arrival that she was, of all things, a doctor of psychiatry.

Did the professor actually believe I was in need of the professional services of a headshrinker? For I could not think anything else of his requesting this Dr. Rogers to meet us. Or was the kindly old man simply trying to fix me up with a date, knowing I was still an unmarried man? Well, no harm done I was thinking. Because it did not seem to matter one way or the other. I say this because when I first beheld the Madonna-like face and curvaceous body of Dr. Rogers, I immediately knew in my love struck heart, I was going to have to fake I was in need of her services. This being necessary if I wanted to see her again.

As the professor and I neared the back of my lab in my Chevy, I noticed by the expression of absolute astonishment on the professor's face, he had suddenly caught sight of my fantastically huge shell. For he then turned instantly to me and said, "My God! Philip! I wasn't prepared for this. That thing can't be real. It's uncanny. Like nothing I've seen. It's simply impossible. Your shell must be a mutation caused by atomic tests because nothing else could explain its size."

"That's it, Professor Moebius!" I shouted. "I

never realized that could be the reason. But that would mean there might be other nautilus shells, as well as other varieties of shells of this size floating around. And God only knows what other living creatures might have been similarly affected."

"Yes, Philip, indeed. Like maybe the mollusk that once lived in that shell."

"I've just given that some thought, Professor. I mean if it was radiation from atomic testing that was responsible for creating that shell's size, then reason has it, the same radiation would have probably killed off the mollusk that inhabited it. I only hope that's also true for the other living things that were exposed to the atomic testing that went on in the oceans of the world."

"I hope so, Philip. Or we'll end up having to kill these mutations with the very bombs that were the cause of their size. It'll become a deadly cycle I'm afraid."

"Hey, is that awestruck woman with the blond hair standing near my shell this Dr. Rogers by chance?" I asked hopefully, as we arrived in my Chevy at my lab. "She sure is a beautiful woman, Professor. Just what kind of a doctor is she? I forgot to ask."

"Well I don't know how to put this to you, Philip, but Dr. Rogers is a doctor of medicine. I thought you might need an examination. I think you could use some rest, because you sounded a might overworked when you spoke to me on the phone. And now that I'm here, I know I was right."

"Well I hope she doesn't mind the fish odor in my lab, Professor, or she might need a doctor herself."

As the professor and I then made our way to where Dr. Rogers was still awestruck and standing, I suddenly began feeling quite strange again. And realizing I was getting too close to my shell, Professor Moebius instantly began yelling to Dr. Rogers to come to my aid, as I was in the process of fainting.

"What's the matter with him, Professor Moebius?" I suddenly heard her say, as I was about to pass out.

"I was hoping you could tell me that, Dr. Rogers. But I believe Dr. Grieg is being mentally influenced by this stupendous shell you were so much admiring. It seems to be affecting him. And I have reason to believe it's because the shell is resting directly over a spot that may be giving it a powerful malevolence. I must inform Dr. Grieg's assistants to immediately move this shell to some other spot."

I then replied, half conscious of my surroundings, "You might be right, Professor. But I think it's only Dr. Rogers' beauty causing me to be so thunderstruck."

"Why, thank you, Dr. Grieg. But I don't think even Helen of Troy's face could have caused you to react this way."

I heard her reply, as she and the professor began helping me into the front door of my lab and to a chair, where quickly regaining my senses, I began saying, half jokingly, "Maybe I'm going to have to wear that Faraday cage after all, Professor," wondering at the same time if I had subconsciously faked my fainting spell in order to gain Dr. Rogers' sympathy, or nearly fainting because of my shell out there.

Funny though, but I had never experienced anything like this before when I had been near a beautiful woman. But then suddenly remembering the affect Whitman's poem had had upon me, as well as the shell's new location as the probable cause of my lightheadedness, I heard Dr. Rogers remark, "What's that? What on earth is a Faraday cage, if you'll both pardon my ignorance?"

"Well let me try to explain my theory about that, Dr. Rogers."

"Please do, Professor Moebius, for if I'm to help Dr. Grieg, then I'm going to have to have a lot more information than you gave me in that phone call this morning."

"Now wait just a minute, the both of you. I don't think I'm in need of a doctor. I mean, just because I thought I heard a voice again and had a little fainting spell is no reason for me to undergo an examination," I said, completely forgetting I wanted to become Dr. Rogers' patient.

"That might be true, Philip," I heard the professor say, "but I'm thinking of all the other times over the years you've told me about feeling strange."

"OK, Professor, I guess you were right about sending for Dr. Rogers as you did," I said, as I then turned to look into her lovely blue eyes.

"Fine then," she said. "And now that that's settled, what were you going to tell me about your theory, Professor Moebius?"

"Only that I believe that unfathomable shell out there must be sitting over some kind of underground spring or watercourse. And if that's the case, then this could be causing all sorts of strange things to result..."

"Like what, for instance, Professor Moebius?" Dr. Rogers said looking like a woman who was now inwardly scared.

"Well, I see a very possible connection with that shell, if I'm correct about where it's now resting, with the ancient practice of locating sacred buildings directly above springs and watercourses, as was done with Solomon's Temple in Jerusalem, and the magnificent Chartres Cathedral in France. The builders of these two holy places of worship must have done this for reasons of a spiritual nature we are at a loss to explain. But I believe I may have finally unraveled the perplexing mystery: When we know that the less conductively a given terrain has, the better will waves penetrate it, it is clear that if we want to have a healthy house we must build it on ground permeable to harmful waves (that is, dielectric), which will absorb rays to a very great depth without causing a reaction on the surface field. To avoid all error, let me point out that these terrains of low harmful density are composed of Stampian sand, limestone, gypsum, sandstone, primitive crystalline rocks and certain recent alluviums rich in sands and gravels.

"Terrains that must be regarded with suspicion, or even considered unsuitable for any perfectly healthy building, are impermeable and therefore conduct harmful waves. They are composed of plastic clay, gypsum marls, Jurassic marls, phosphated chalk, iron pyrites, carboniferous soils, and iron-ore shales.

"With a permeable terrain, the surface field is not altered; with an impermeable terrain, 'reflected, refracted, and diffused rays are compounded with the original rays to produce an interfering field and standing waves.'"

"That's quite a mouthful, Professor. You should have become a geologist. But I'm not too sure I fully understand what you've been saying," I heard Dr. Rogers say, obviously having a hard time dealing with the professor's scholarly discourse, as I quickly added:

"That goes for me too, Professor."

"Then let me see if I can simplify what I've been saying to you both...It is possible that carrier waves coming from the underground current have been charged far upstream, with abominable emanations that, modulated by a magic frequency, are directed to the bedroom, which acts as a resonance chamber...Of course, when I say 'bedroom,' I really mean that monstrous shell out there in your particular case Philip, and not in the room you first caught sight of it acting as this 'resonance chamber'... What it all boils down to is this: When a telluric current, forming a horizontal electric field at the surface of the earth, meets a magnetic field formed by vertical cosmic rays, a very powerful and dangerous wave is produced at the intersection...The fault lies with the water: it transports harmful waves emitted by minerals or rocks over which it has passed."

"But Professor, why aren't any of my assistants, Dr. Rogers or you, being affected by these carrier waves? If that's the reason why I'm feeling so depressed now."

"I didn't know you were feeling that way, Philip," the professor was telling me with deep concern on his face. "But it seems some people: 'Are psychomotor centers highly vulnerable to emissions of abstract waves. It is possible that they unknowingly serve as relays and transmit the emissions they receive in the direction suggested by them…' This, of course, would in return make you feel in a depressed state of mind, a condition which Dr. Rogers should be able to quickly remedy. So I would suggest before Dr. Rogers examines you, we first have some of your assistants move your shell to another location immediately. This should prevent you from having some sort of a nervous breakdown."

"I can fully concur with that advice, Professor Moebius," I heard Dr. Rogers say, as I nodded my head and added:

"But wait a minute. I think there might be an easier way to prevent these mind-altering waves from using my brain as a kind of transformer. And I don't mean by wearing a ridiculous Faraday cage over my head, as you earlier suggested Professor."

"Well, Philip, if you've a better idea, tell me, because anything's got to be easier than trying to move that mind-boggling shell to another spot. So what exactly do you have in mind, my dear boy?"

"Well, what if we first turn the opening of the shell directly away from my lab. I say this because I seem to remember concave objects, such as pots, or hats, are able to emit malignant waves. For example: People and dog attacks…If they are wearing a bowler hat, or can pick up something like an old pot, they should turn the concavity toward the attacking dog, so that he sees its bottom; he will bark furiously, but he will back away, not daring to attack a person protected by the wave of a concave shape..."

"Yes, Philip, that's a good idea about repositioning the shell's opening. But if that's all you're going to do, I don't think it'll be enough to stop the shell's affect on you."

"Yes, I suspect that will be the case, Professor. But what if we then make some kind of a protective and insulating barrier around the shell by driving small wooden stakes into the ground, then connect all of them with electrolytic copper wire that's been coiled around them? Wouldn't that then insulate the shell from energies arriving from outside the shell's immediate area?"

"Yes, that might. However, the end of this wire would have to be grounded far enough up stream, or away from this spot. But then what? I mean what are you going to do then? Admire your beautiful shell? Because even though your shell is one hell of a thing to gaze at and admire, it's bound to have an affect on your health, one way or the other. Then think of all the people your shell will most assuredly attract. You wouldn't be able to get any work done anymore."

"I agree with Professor Moebius one hundred percent, Dr. Grieg. But why don't you give your shell to a museum, or to my university? I'm pretty sure the biology department would give just about anything to get their hands on it."

"I'm afraid that will be out of the question, Dr. Rogers," the professor was saying. "The shell is simply too dangerous to be placed anywhere where people would be exposed to its harmful influence. At least not until we know for certain how it can be rendered completely harmless. And besides, I think I would like to try a little experiment with that beautiful and mysterious oddity."

"Just what do you have in mind, Professor? I'm almost too afraid to ask."

"Well, Philip, it has to do with the Moebius strip, if you haven't already guessed. But I'm wondering if there might be another danger far greater within your shell we have as yet to discover."

"Like what, Professor Moebius?" Dr. Rogers said, as she then looked at me with concern.

"Well, we all know the paradox of the Moebius strip, which has only a single surface, possessing neither a front nor a back. Well, have either of you heard of 'Jordan's Curve Theorem', which postulates the idea that our universe is similar to a closed curve, much the same as the Moebius strip? Well, I was thinking there might be an 'inside' and an 'outside' universe coexisting in the same space with our own. In other words, there may be two regions of the space-time continuum penetrating each other despite their objective remoteness. So it may be possible then to bring about a sudden passage, if you will, into another time or another space."

"You mean by using my shell as this passage you're theorizing exists, Professor?"

"Yes, Philip, exactly, but we would first have to construct a giant metal Moebius strip, and then position it inside your shell. Then I would have to take a little ride on it."

"What?" I said loudly, "You can't be serious, Professor? God only knows what might happen to you in there," I finished saying to him, obviously shocked by his idea.

"Yes. That's very true, Philip. But that's going to depend on whether we put a half twist, or a full twist into our metal Moebius strip."

"You know, Professor Moebius, only an astrophysicist like yourself could have come up with such a theory."

"Why thank you for the compliment, Dr. Rogers," I heard the professor say with a chuckle. He then turned to me smiling and said,

"So what do you say Philip? Do you want to try your luck with me in trying to unravel the secret of my great-grandfather's discovery?"

"Sure, Professor. However, you're going to have to promise me you'll stop your experiment if anything out of the ordinary occurs."

"But that's exactly what we'll be trying to achieve, my dear boy. And besides, I'm an old man of eighty now. And I'll be damned if I'm going to my grave without finally unraveling the secret of the Moebius strip now that an opportunity like this is presenting itself."

"All right, Professor, let no man say I dared deny the great-grandson of Dr. August Ferdinand Moebius from fulfilling his destiny," I said to him warmly with a smile, as I then noticed Dr. Rogers looking at me with what I knew to be a flicker of love in her eyes, the poetic words of Whitman's *Leaves of Grass* then flooding my mind:

Behold a woman!
She looks out from her Quaker
Cap ... Her face is clearer and
More beautiful than the sky.

Chapter Four

It had taken nearly an hour to turn the shell totally around, so its ten-foot opening, that had been facing my lab, was now pointing in the opposite direction.

My five lab assistants had done the actual job of placing the wooden stakes coiled with electrolytic copper wire into the ground, completely encircling the shell at three-foot intervals. I had not attempted to help them because of the shell's obvious affect on my mind.

But no sooner had they completed the job, than I observed Professor Moebius, suddenly, and without as much as a word, begin to calmly walk toward the shell and just as calmly enter into its opening.

He had been in the shell for less than ten seconds when to Dr. Rogers' and my surprise, we saw him come running wildly out of the shell, yelling at the top of his lungs, "My God! My God!"

And as Dr. Rogers and I watched with great fear and concern the professor then come to a sudden halt, some fifty feet behind the shell and its oval-like opening as if totally frozen with fear, we immediately began racing out of the lab, from where we had been watching the whole disturbing affair.

On finally reaching his immobilized body, Dr. Rogers and I noticed he was mumbling something as we tried to console him.

"What's wrong with you, Professor? What happened to you in there?" I said with a grave concern, noticing an expression of what I knew to be horror and shock on his wrinkled face.

"I heard the voice, Philip! I heard the voice! It talked to me! God! You were telling the truth!"

I knew by the way the professor had just answered me that he was truly scared out of his wits, for he had spoken to me with little shrugs, as if readjusting his vertebrae.

"What did the voice tell you, Professor? I said, with a cold chill in my heart.

"The voice called your name, Philip! It called your name!"

"My God!" I said, taking a deep breath, like I was about to be shot.

"Is that what the voice in your mind said to you this morning, Dr. Grieg?"

"Not exactly, Dr. Rogers. It was more of a feeling of danger and dread than anything like a voice," I said, still wishing I had actually heard the voice a second time in my life only hours before.

"But, Professor Moebius," Dr. Rogers said, turning from me, "was the voice you heard in the shell that of a woman?"

"Why yes! It was most definitely a woman's voice I heard, but also more than just a voice. Something I'm too scared at the moment to explain. And even if I wasn't a nervous wreck right now, I don't think I would want to relive in my mind the experience for you both, at least not while I'm fully conscious to hear my old trembling voice."

"Of course, Professor. We understand," I said. "And I think it would be a good idea if you went into the lab now and rested."

"Dr. Grieg is right, Professor Moebius. You can always tell us what happened to you in there tomorrow. And if you want to tell us your experience under hypnosis, then that will be fine. But I must say, it was very foolish of you to have gone into that shell, especially after what you previously told Dr. Grieg and I"

"Yes, I know, Dr. Rogers, but I just couldn't resist the urge. Maybe that's because we scientists can't help but want to investigate things. And besides, I thought it would be safe to go inside after we had gone to the trouble of grounding it."

"That's true, Professor. But we didn't ground the inside of the shell, only the ground around it, so you had no business going into the shell until after we first tested it was safe."

"I'm really sorry, Philip. I guess I wasn't thinking. I was just too eager about seeing what it looked like inside, but I guess I'd better go rest now. My nerves feel really shot."

As Dr. Rogers and I watched with concern the professor head toward the lab, half dazed and refusing our help, I grabbed her by the arm and said, "I've never seen the professor so scared. I'm wishing I'd never phoned him because now I'm going to have to go into that shell again, whether I like it or not. But I can't understand why the shell acted upon us both in almost the same manner? And why and how in God's name did he say the shell called out my name to him? Because we can't both be suffering from a case of paranoid schizophrenia."

"Dr. Grieg, you aren't seriously going to go inside the shell after what just happened?"

"Well I can't think of anything else that's going to stop this fear and depression I've been having all day. I mean, if I don't face up to that shell and whatever's inside it, then I'm really going to go mad. After all, wasn't it President Roosevelt who said, 'We have nothing to fear, but fear itself'?"

"I admire your courage, Dr. Grieg. But let me inform you, I'm also a doctor of psychiatry, as well as a doctor of medicine."

"I sort of gathered you were. I mean by the way you acted when we first met. It was the sympathetic understanding in your voice that gave you away, Dr. Rogers."

"I think we had better call each other by our first

names. There are many doctors around here, what with your assistants. My first name is Elaine."

"Beautiful name, Elaine... Hey that rhymes," I said.

"Don't tell me you also write poems, Philip?"

"I wish I did, Elaine. Still, I do love reading them, especially the poems of Walt Whitman."

"Oh he was a great poet, but hard to understand. I mean he was so ahead of his time. Wouldn't you say?"

"Yes. That's true. But he was also a kind of a mystic."

"I'm aware of that, Philip. I happen to have studied his poems at college, as well as the ancient poetry of Ireland,"

"That's a surprise. You'll have to tell me about that some time, Elaine," I said, quite interested.

"I may have to tell you about it now, Philip."

"Why's that?" I said, half dreading her reply.

"Well, I think I might possibly know the reason to why both you and Professor Moebius heard a female voice in that shell of yours. But as of this moment, it's only a guess."

"Well fill me in on it before I go into the shell and get my brain fried, most likely, because you just might be able to dissuade me."

"I think Professor Moebius should be present when I explain my theory, Philip. In the mean-time, I can tell you the probable cause of your depression if you like?"

"OK, shoot. You might as well give me your version also, Elaine."

"Well it may be that the shell is blowing out from its opening a strong current of positive ions, or what in nature is called a 'Witches Wind'. This wind of positive ions is also called the 'Foehn' in Austria and is known to be responsible for making people depressed enough to commit suicide. Those people that is who happen to live in the region of Austria where this 'Witches Wind' is blowing. This region is even referred to as the 'Suicide Belt' and it stretches up through Middle Europe. The suicide rate soars whenever this wind is very active. As a matter of fact, there's such a wind of positive ions coming from the region of Santa Ana, California. This could be the reason for your feeling of depression, taking into account the present location of your shell and the wind currents of the 'Jet Stream' - the shell may be acting as a sort of a funnel and wind tunnel. Then again, maybe the professor's theory is the correct answer to what's happened to you both. Or it may be both of our theories are wrong, the two of you only suffering temporarily from schizophrenia, caused by something as simple as a brain chemistry change being brought on by something you both may have eaten and inhaled this morning."

"But, Elaine, how can a wind of positive ions cause a person to commit suicide?" I asked, as if my life depended on her reply.

"Very simple, Philip. You see, humankind evolved in conditions of normal ionization, making us, as a result, bioelectric beings created by God to live in an environment containing a certain level of air electricity. However, when there are too many of these positive ions in the air, your body starts producing an overdose of the stress-response nerohormone - 'serotonin.' This in turn can upset your body chemistry causing depression, despair, and anxiety; but only to some people, who like yourself and Professor Moebius, it seems, are acutely sensitive to positive ion poisoning. So I would not advise going into that shell under any circumstances until you can figure out a way to avoid breathing into your lungs these positive ions, or getting them on your body for that matter. They tend to be attracted to body hair and peoples' clothes."

"You know, that might explain why the professor came out of the shell looking so scared... Wait a second, but if these positive ions are above normal in the shell, then wouldn't it cause him to react in a suicidal manner after inhaling them?"

Looking at each other as though we knew we had made a terrible mistake letting the professor go off alone into the lab, Elaine and I both turned in panic in the lab's direction. It was then we saw through one of the lab's windows, to our horror, the professor with a knife in his hand.

So racing back into the lab as fast as we had raced out after seeing the professor come yelling out of the shell, we were again shocked to see him about to slash himself across his wrist with the knife.

"Professor Moebius!" I yelled, as I quickly ran to him and grabbed the knife from out of his hand. "What the devil were you trying to do? Kill yourself?" I said to him, noticing the confused look in his eyes, looking at me as though I were a stranger. "Elaine, what's wrong with him? He doesn't seem to know me."

"I'm not really sure. But it looks like he's under a spell of some kind."

"My God! You're not thinking what I'm thinking, are you Elaine?"

"Yes, Philip. I'm afraid I am. So we're going to have to un-hypnotize him from whatever has a hold on his mind... But lay him down on that sleeping cot over there first, so I can give him an injection to calm him."

Doing as Elaine had instructed, I slipped the professor's right arm over my shoulder and guided him to the cot I used for naps. I was beginning at this moment to fear for our lives. What the hell was that shell doing to the professor and me? I was in mortal fear and wondering. Would Elaine be the next to be affected? That dreadful thought by itself almost caused me to go out and dynamite that haunted hell of a shell. But I did not have the heart to do it then. The shell was just too beautiful to destroy. Yet I knew at that moment I was perhaps dealing with something in that shell from some other world. And the fact I was a marine biologist made it imperative I discover exactly what that something was, as well as what was causing the shell to behave in this diabolical way.

And if there was, in some strange manner, a spirit in the shell now making it its home, which I was beginning to suspect, then so be it, for I could always summon a priest to perform the required exorcism. Yet, what if we were dealing with something nobody knew a damn thing about - including the priests of the clergy? But then I suddenly remembered what Elaine had said about having a theory about the voice.

"OK, Philip, you can roll up his sleeve now," I heard Elaine say, as I did what she asked and watched her give the professor an injection of Valium which she had gotten from her car. And as I saw the professor suddenly regain his senses seconds later, coming out from whatever had had a hold on his mind, I wondered if perhaps the shock of the needle itself had been the cause.

"Professor Moebius? Are you all right? What was it? Do you recognize me?" I then asked, half expecting him to return to the semiconscious state he had been in, then waiting for him with a bated breath to respond to my words.

"Yes, Philip. I'm all right now. But don't allow me to be left alone again because that shell seems to be affecting me as well. We must have done something to make it more powerful. It must be because of the grounding. We'll have to get your assistants to change the electrical polarity of the ground surrounding the shell... No! That won't do! We'll have to seal shut the opening of the shell with a metal door... Yes! That should prevent any harmful emanations from escaping."

"But, Professor, why did you just try to kill yourself?" I said with an apprehensive fear, wondering if Elaine's theory of positive ion poisoning had made him want to commit suicide, or if he even remembered his failed attempt.

"I really don't know, Philip. And it doesn't matter to me anyway. And I know that sounds pretty strange. But what's important at the moment is for us to get a metal door on that shell, and then go ahead with my experiment with the strip immediately."

"But aren't you afraid something else might happen to you if you re-enter my shell again?" I said. "Because we still don't know what we're dealing with in there. And I'm already regretting I got you involved in this. If anything happens to you I'll never forgive myself."

"It doesn't matter what happens to me, Philip. I'm too old to worry. What matters is finding out what will happen when we place a Moebius strip inside that shell of yours. Then like I told you, I plan on taking a ride on it."

After having heard Professor Moebius talk about his imaginative and daring plan again, I knew he was either mad as a hatter, or being the scientist that he was, could never be blamed for wanting to risk his life in an experiment we both knew was insane.

Yet why had he been so unconcerned with the fact he had just tried to kill himself? Maybe Elaine could explain that to me? I wondered at the time. But how was I going to talk to her in private now that none of us could be left alone near the shell? I guess I was going to have to call her at her university, or slip her a note in the lab before she left. Now that I knew the shell was a danger, I was not about to let the professor or the shell out of my sight, because he would probably want to live here at my lab instead of going back to my house on the island. And that meant I was going to have to keep a watchful eye on him constantly, so going back to my house without him was out of the question. But then I said to Elaine, who had been silently observing him.

"Elaine, I think it's the right moment to tell Professor Moebius and I about that theory of yours. The one about the voice I mean. Because if there's anything to it, then we might be able to proceed with the professor's experiment with at least an idea of what we're up against."

"What theory?" I heard the professor ask, with a quiver in his voice.

"Oh, while Dr. Rogers and I were outside talking, she told me she might have the answer we've been looking for. Something to do with Irish poetry if I remember?" I said with a smile as I glanced her way.

"Irish poetry?" the professor said, his bushy white eyebrows rising, I observed.

"Yes, I know it sounds strange, Professor Moebius, to say the least," Elaine was saying, "but most of the Irish poets of the Middle Ages all seemed to be writing about something that might well have a connection with what's been going on here; something the great psychiatrist, Carl Jung, referred to as the, 'Anima', or the personification of the feminine attributes existing in the minds of men."

"But what does Irish poetry, or this 'Anima' thing have to do with Dr. Grieg's shell, Dr. Rogers?"

"Well it seems these Irish poets were all writing about this 'Anima' Jung talked about, but were too ignorant to know it at the time. They were instead all writing about the external manifestation of the 'Anima', or what they all believed to be 'The Earth Mother Goddess', who in Greek Mythology, and other cultures, is called by different names."

"You mean like, 'Aphrodite', 'Athena', or 'Isis?"

"Yes exactly, Philip. But there were other names by which people called her, such as: 'Diana', 'Cybele', 'Lilith', 'Medea', 'Nanna', and 'Kali', as well as: 'Helle', 'Salmaone', 'Sarah', 'Tamar', 'Acca', 'Aegea', 'Anu', 'and 'Astarte'. And these aren't even the Irish ones."

"Hey wait a second, Elaine, that might explain why you haven't been affected by the voice as yet," I said.

"You mean because of the fact it was a woman's voice you both heard?"

"Yes exactly," I said.

"Well according to Dr. Jung, each woman's psyche possesses a corresponding personification of her repressed male attributes. This, Dr. Jung named: the 'Animus'. So you may be right, Philip, about why I haven't as yet been mentally influenced or affected by her voice; a voice I now more than ever suspect to be coming from the 'Earth Mother Goddess' herself. But it would seem she only wants to make herself known to you and Professor Moebius, and not to the other men working here."

"But how can you be sure it's this goddess?" I said. "Because we still don't know if the shell's being acted upon by some sort of ion or magnetic anomaly, or what's causing the professor and I to feel depressed and hear voices. And besides, how in hell did this mythical being, or whatever these poets called her, get into that shell out there? Which I might add, you can't seriously think Professor Moebius and I are going to believe even exists, disregarding Botticelli's *Venus on a Half Shell* painting as your proof."

"Of course not, Philip. At least not in the way you just said. Because you're forgetting, I said this being, or goddess, the ancients believed to be a real person when they saw her appear before them, was only the repressed feminine attributes being projected from out of their unconscious minds."

"Wait a second. Did you say, 'before them', Dr. Rogers? Are we to believe these poets actually saw this 'Mother Goddess' you're talking about?"

"Yes, I am, Professor. And I'm not talking about them seeing little green fairies, or seeing something of that sort because of getting drunk with strong drink. Because it was almost always in the sky above them that they saw this divine figure of the 'Earth Mother

Goddess'."

"You mean like at Fatima, Portugal, in 1917, where seventy thousand people were present when three children saw the figure of the Virgin Mary in the noon day sky? Or recently at Zeitoun, in Cairo, Egypt, at a Coptic Orthodox church in 1968 through 1971 where hundreds of thousands of people were said to have seen her as well?"

"Yes, Philip, exactly, but those aren't the only places the 'Earth Mother Goddess' has appeared to mankind. There are countless other places around the world I could site."

"Then please do. We need to know as much as possible about this strange phenomenon. Because for all we know, we may be involved with something that could save this planet from some future threat. I mean if we could find a way to communicate with this feminine unconscious projection, and then somehow convince her to appear over the Soviet Union, she might be able to convert that Godless Empire to Christianity. And just think what that could mean to the world?"

"Yes, that would be wonderful, Philip. But it's too early to start speculating about what the true cause of this voice is because I'm only guessing about the whole thing anyway. So I can only suggest the both of you proceed with as much caution as possible."

"Yes, Elaine, that's sound advice," I said. "But I hope our experiment doesn't end up releasing something we'll all be sorry for, that shell could be another 'Pandora's Box' containing some deadly evil for all we know."

"Yes, that's a possibility," the professor said, then added, "But it may be the only hope for the world... I mean what the hell does it matter if the devil himself should happen to come out from that shell because of our tests? The world will probably blow itself up in a few years anyway. But if we can make contact with this 'Mother Goddess' of yours, Dr. Rogers, be her the 'Virgin Mary' for that matter, we can only pray, we might very well be able to prevent another world war from occurring. And that's enough of a good reason to go ahead with my plans at any costs."

After having heard Professor Moebius speak, I knew he was 100% sane. It was as if he had at long last discovered a secret purpose for his life, linked together in some mysterious way, with my shell and his great-grandfather's discovery.

Chapter Five

It took almost an hour for my assistants to secure an outside iron door to my shell's opening, as well as to string a line of electric lights inside of its iridescent white 'mother of pearl' interior - almost as long as it had taken them to move the shell's opening away from the lab twenty-four hours earlier. This had immediately caused the professor and I to feel instantly at ease with ourselves - the fear and depression we had shared finally disappearing from our minds.

My assistants, it seemed, had not been affected by the shell's mind-altering effects. And it was obvious the shell was in perfect harmony with the electromagnetic field, or aura, surrounding only the professor and me. This interesting fact had been explained to us by Elaine only an hour before the iron door had been positioned over the shell's oval entranceway. Elaine had referred to a book she had been reading on the subject of electromagnetism she had checked out from her university's library. She had informed us, that because both the professor and I had experienced unconscious feelings of fear and suffering, we may have .given rise to exceptionally intense electromagnetic waves, thereby causing the shell to fix these waves inside its hollow chambers, and reconstitute them when they encountered a suitable environment - the waves then converted themselves into sound or light by a system analogous to that of the feather of a bird, particularly the feather of an owl. According to Elaine, the ancient Greeks considered the owl as their symbol for the goddess Athena. Why this was so was anybody's guess. But in this century it was discovered birds' feathers could transform the singing of a bird into a train of electromagnetic waves, which could then be propagated at the speed of light. And it was similarly discovered that human hair, including whiskers, could function as a microphone and transmitter, thereby enabling the professor and I to pick up in our minds any electromagnetic waves being transmitted in the air in direct proximity to the shell.

Yet who knew for sure if this was the correct explanation for what had caused the voice. But then Elaine had gone on to say that everything had a voice, that all creatures were eternally singing the praises of the Creator.

Still, all this I had already known. For the professor had explained this to me many years earlier, but in more of a scientific way, saying in effect, atoms could produce sounds by their reverberations.

Yet what in the long run did these explanations and endless theories really matter, they could never explain away why we all seemed to be so scared. All the same, we knew we had to go on with the experiment. Whatever the true explanation of the voice, we knew we had to find this goddess, if that was the true thing Elaine had explained was in the shell, being projected into it by either the professor's or my mind, then using our brains like some sort of radio unintentionally perhaps. And secondly, we had to go on with the experiment before the professor or I committed suicide unexpectedly, caused by our depressed states of mind reestablishing themselves.

Still, was it any wonder then, I had been so captivated by that voice? A voice I now longed to hear again, even though I was still scared to death of its origin. But then I was reminded it was not nice to fool with Mother Nature or God's heavenly laws. And I began praying this experiment would not anger whatever inhabited the shell, thereby causing it, or God, to punish us for our foolish tampering, because at the moment I was beginning to feel like Alice in

Wonderland. I was hoping when I entered the shell again, like I knew I would be doing shortly, I would not discover a human-sized rabbit waiting there to transport me to some other realm.

By the next morning, however, while the professor and I were having breakfast at the lab waiting for Elaine to arrive, I had my five assistants begin constructing the giant metal Moebius strip railroad style tracks incorporating its half twist, making sure the strip conformed to the shell's spiraling interior, the many chambers of which they had already broken into and investigated.

My assistants had earlier asked me, after having constructed the iron door for the shell's entranceway, if they could go into the shell and take the required measurements for the placement of the Moebius strip and lights. But after having seen the professor come running out of the shell yelling at the top of his lungs like a madman, all of them began to reconsider the idea. However, their decision to enter the shell became unanimous when I promised to construct for each of them a Faraday cage to wear over their heads, knowing I could not take the chance any of them would hear the voice and go raving mad. But would the Faraday cages also prevent them from seeing a vision of this goddess? I wondered at the time. But since I had not seen this goddess myself, I did not let that thought bother me.

It was then, however, before I had constructed their Faraday cages, that I had decided to fully inform them of what had happened to the Professor and myself, asking them immediately afterwards if they still wanted to go into the shell to light its interior. Having answered yes to my question, after I had fully explained what had occurred the past twenty-four hours, they all agreed, if going into the shell would further the cause of science, then they would gladly volunteer. Of course they had all said this only after I had promised them their protective helmets to wear. Nevertheless, being all cautious, as I knew them to be, they had decided to draw straws as to who would be the first of them to enter the shell. There was no sense, they had said, for all of them to enter the shell at the same time if their helmets could not do the job of protecting them.

Then again, nobody knew for sure what was going to happen when they all eventually entered that thing. But being the adventurous types I knew them to be: three men and two women in their late twenties, I knew in the long run their desire for exploring the unknown would eventually win out over any of their fears.

Of course, you are probably asking yourself why they had not all gone into the shell when they had first helped me to take it out of my cave. Well, that was because the shell's opening had been filled with so much seaweed and sand it had made this impossible at the time. Plus the fact all nautilus shells have many closed off chambers - chambers of which the baby mollusk had to break into as it grew to adult size.

It was immediately after the giant metal Moebius strip had been placed inside the shell, with no ill affects to my assistants that we all went about the task of constructing the motorized vehicle that would ride upon both the outside, and because of the twist, the inside tracks of the strip. Professor Moebius had instructed my assistants to purchase from a nearby railroad junkyard, thirty-two metal train wheels with inner rims, which we then fitted with outer magnetic linings, so as to enable the wheels to hold all the vehicles to the strip's metal tracks. It was going to be necessary for this four-wheeled vehicle to ride upside down when it turned the half twist that we had designed into the strip. The vehicle and its rider were going to have to ride on the inside surface of the strip, as well as the outside, as they would do at the start of their voyage.

Of course, the rider of the vehicle would be strapped in with seatbelts to prevent him from falling out of the cockpit. But the seatbelts would only be needed to hold the rider in his seat until the vehicle attained a speed producing the centrifugal force to keep the rider in while traveling upside down - much the same with riders in a roller coaster while it did the 'loop to loop'.

The vehicle itself, which was black in color, took only two days for us to build, as did the backup vehicles constructed in case the first one failed. These other vehicles had been constructed for another reason however: to serve as rescue vehicles should the first one vanish into some unexpected dimension - the possibility of that happening being very remote in our opinions, but better to be safe than sorry we had said.

All of the vehicles had the appearance of short metal hot dogs, so Elaine commented, as she curiously watched the whole goings on the four days she was present. And it was only after the construction of the vehicles that I sensed Elaine knew in her heart I would be taking the professor's place as the vehicle's rider. Even though the professor had said he was going to be the one to take the first ride, I knew he still entertained a fear of hearing that frightening voice again. And Elaine and I knew his old heart would never be able to withstand the strain his eighty-year-old body would have to endure when he traveled upside down on the strip at the required speed.

Moreover, the voice in the shell he had heard had called my name. So with these facts explained to him he reluctantly gave in and agreed I should be the rider. As for what was going to happen to me while I rode inside the vehicle, well that was purely theoretical.

Nevertheless, it was about three hours before my dare-devilish trek, that I began asking the professor and Elaine what I might expect to see and hear when I entered the shell this second time in my life. They could only give me many explanations again, with neither of them positive about anything, as I should have expected.

Then finally, as the inevitable moment arrived for my momentous trip, I suddenly began to mentally sense the metal door of the shell being opened by my assistants as I waited nervously in my lab. An ice-cold chill then began to run up the length of my spine as I immediately reacted with a feeling of fear I knew would remain with me for some time, knowing also, I would not be wearing a Faraday cage over my head when I went into the shell. This decision I had agreed to in order to experience everything that was to happen while I was inside the shell and speeding vehicle. Any danger, either mentally or physically, that might happen to me while in the shell, I was immediately to respond to by pushing a panic button on a device I would be wearing. The professor and Elaine would then react to my call by sending my assistants into the shell to take me out if the need turned out to be warranted.

Even with a panic button at my disposal, I was not about to chicken out, no matter what I might happen to see or hear while on my ride, at least that is what I was telling myself. And besides, my mental and bodily reactions would be closely monitored on both a brainwave machine and a video camera - the camera having been mounted at the front of the vehicle allowing Elaine and the professor to observe my every move. Even my heart rate would be recorded for future study during my harrowing ride. And all these added precautions leading up to my little jaunt into who knew where was only making me feel as though I were about to travel to the moon in a spaceship or something. Yet in my particular case, I wondered if I would return from my voyage a freaked-out brainless vegetable, or for that matter, come back at all.

As I finished watching my assistants push the tubular shaped vehicle, which was traveling on wooden railroad tracks, into the opened iron door of the shell, and then lift the vehicle's magnetic wheels upon the metal tracks of the strip by way of a ramp; I suddenly turned to Elaine and kissed her on the lips. She did not resist.

"Oh, Philip," I heard her say after we kissed, her voice delightfully swooning, "I wish you weren't going on this crazy ride of yours because I'm falling in love with you."

"Don't worry, Elaine. I'll be OK. Besides, if I should vanish in there, you could always come rescue me."

"Please, Philip, don't joke. There's no telling what might happen to you. You might enter another dimension, as the professor warned."

After Elaine had spoken, I knew Professor Moebius had heard, for his eyes now had a grave and worried look. "Don't worry, Philip," I heard him say. "If you do enter another dimension, which seems slim, I could always get you back by reversing the direction of the magnetic field surrounding the shell. Or if that happened to fail, I could put a full twist into the strip instead of the half twist it already has. That might interlock our three dimensions with this fourth one preventing you from being locked out from our space-time continuum. Then I, or perhaps Elaine, would have to follow you into this other world and rescue you."

"Oh that sounds peachy, Professor," I said half tauntingly, "But what if I'm dead after entering that place, or a total mixed up mess from having my molecules rearranged? Not that I'm worried enough to back out of going on with this experiment at this late stage. But did you ever give serious thought as to what we might be doing to the nature of the universe by this experiment of yours? Because for all we know we might be doing it great harm."

"Yes, Philip, I've pondered this over in my mind these past four days, but like we've talked over, the world is heading for disaster anyway, what with the 'Green House Effect', the over population, the coming food shortages, and the predicted Armageddon in the Middle East. However, if we can save this planet by making contact with whatever's in that shell, as crazy as it sounds, then it's worth the risk to go ahead with this, whatever the costs."

"I guess you're right, Professor. We can't think of ourselves. We have to think of this planet's future, even if the experiment ends up killing us both."

Having said that, I turned to Elaine and remarked, "That's why the professor and I became scientists. To risk our lives if ever the work warranted it. And I guess this is one such case in our careers, I'm almost sorry to say."

"And I'm sorry we had to meet each other under such circumstances, Philip. If only we could have met later, after this whole affair, because parting is such sweet sorrow and more than I can stand."

After kissing Elaine a second time, I turned to the Professor and shook his hand. Then exiting the lab, I entered the shell in almost a panic and immediately sat down in the vehicle's seat. As my assistants lowered the glass canopy over my head, after having connected me up to the monitoring devices inside, enabling Elaine and the professor to receive all the information on my mental and bodily senses, I suddenly realized I had not heard the voice in the short time I had been in the shell. Something had prevented me from hearing the voice and I had no notion why. Maybe I would hear the voice in the next few minutes. I was wondering, and maybe not. But what did it matter anyway, I had a job to do; a job that might in itself cause this voice to make itself known to my consciousness again. Wasn't that exactly what the professor's experiment was really about? Maybe this being that the voice belonged to was purposely waiting for me to reveal my intentions in the shell before appearing to me.

I then remembered I had not looked back at Elaine or the professor while I had made my way into the shell. I feared this would only have caused me to question my sanity about risking my life. But the fact that the iron door of the shell was going to be left open during my ride was something of a comforting thought. We had reasoned, with the door left open, not only would I be able to get enough air to breathe, but any radio transmissions I would either transmit or receive in the vehicle would be able to pass through this opening without any interference. The strong field of electro-magnetism still surrounding the shell and coming from the magnetic underground currents, we realized, might still be a problem though - as far as the radio being at all operable was concerned. However, it was this surrounding field of magnetism we hoped would cause the vehicle and I to react in a most unusual way as we traveled around the strip. The professor believed this field of magnetism would cause the shell to react on the vehicle to produce what he called: 'a spherical resonance'. This in turn, the professor hoped, would cause harmonic vibrations that would then combine with the sounds already echoing in the shell. This he believed would then alter matter long enough to cause - 'a rhythmic articulation of time and space'. This theory the professor had learned from his exhaustive study of vibratory physics and electro-acoustics. He also believed there would be produced in my mind a harmonic coupling of my brain neurons with that of the earth's planetary resonance. If all this actually occurred, it would then end up causing me and the vehicle to enter some other dimension the professor so fervently hoped.

Of course Professor Moebius had assured me the possibility of entering this other dimension was, if anything, more than remote, for what we were attempting was the first experiment of its kind, so he did not know if only my mind would be affected, or only the vehicle.

Would I perhaps only experience a vision? Or would I have my spirit together with my soul catapulted out of my body to who knew where? We all wondered. And to say I had not been scared out of my wits by all these possibilities would have been an understatement. But at least I now had a good enough reason to want to risk my life on this topsy-turvy ride coming up.

Did the shell in any way have a connection with England's 'Stonehenge'? I then began to ponder. Because there seemed to exist among the Masons of the Middle Ages a strong held belief in the mysterious relationship between sound and architecture. The Professor had only confirmed this belief in me when he had earlier brought up the fact that all the famous churches of Europe were built over places similar to where my shell was now resting. It was these areas of magnetic anomaly that had caused birds to build their nests nearby. Had this been the reason the ancient Greeks used the owl as their symbol for the goddess Athena? I had reasoned at the time.

Then as I checked the radio in the cockpit, I began to marvel at the simplicity of the vehicle's construction. The hull was composed of galvanized steel, so thin it almost reminded me of aluminum foil. The vehicle had to be light enough in weight, as well as electrically magnetic, so as to prevent it from falling off the strip as it flipped upside down because of the twist. The means of propulsion we had incorporated into the vehicle was a regular four-cylinder gasoline engine. This we had determined to be the only possible way to get the speed we knew would be needed. And this was the third reason the iron door of the shell had to be open - to prevent me from dying of carbon monoxide poisoning, plus the fact I had to see where I was going in the shell in the event exhaust from the engine caused any blinding black clouds. The velocity of the vehicle was something of a mystery to us though, for we did not know exactly what speed the vehicle would have to attain in order to reach the same resonance or frequency, as that of the molecular structure of the shell. Because it was this that was to produce what we wanted: a rhythmic articulation of time and space, thereby changing my reality as I knew it, as I raced around the twist of the strip.

So buckling up my seat belt and shoulder harness, I said into the mike of my radio, "OK, Professor, I'm going to start up the engine now."

"OK, Philip. But be careful you don't go too fast at first; we only want to test out the vehicle this first time. So if anything out of the ordinary starts occurring, then proceed with care."

"But, Professor, what if I should lose consciousness during the ride?" I said with some dread.

"Then we'll just have to let the vehicle run out of gas, because I can't stop the vehicle. But since we only put a quart of gas into the tank, we don't have to worry too long."

"Philip, I love you," I then heard Elaine say warmly, overjoyed by her vow.

"I love you too, Elaine," I answered as warmly.

"I love you too, Philip," the professor said next.

"I didn't know you cared, Professor," I replied with a laugh, then added, "I'm only joking," as I revved up the engine placing its gear into drive and said, "The engine sounds great, Professor."

"Keep your speed under twenty, until you get a good feel for the vehicle," he answered.

"Will do, Professor," I said.

As the vehicle started to move now, I suddenly began feeling a bit strange again. And the fact I had not heard the voice in the shell as yet, was more than adding to my concern. Would the voice call out to me as it had done to the professor? I was wondering. Or would I hear the voice a hundred times more amplified because of the echoing effect of the glass canopy now over my head like a punch bowl?

Before I had the time to frame my next thought, it began to happen: faint, shifting notes of song, almost too high in pitch for my bedazzled ears to fathom, but for a slight reverberation echoing throughout my body. I was wondering if perhaps this was being caused by the quantum responsiveness of my body's atoms.

As the vehicle and I began to pick up more speed and then suddenly flip over because of the twist, I couldn't help but be reminded of the relativistic limerick:

There was a fast driver named Dwight,
Who could race at the top speed of light,
He would take off in a flash nearly crap
In his pants, but come back on the previous night.

Then suddenly, and to my surprise, I seemed to be able to remember what I imagined to be some of my past incarnations, at first only a few. But then like Pythagoras - who I knew had only been able to remember twenty of his past previous lives, I was instantly remembering dozens of my own.

I was at that moment more than a little tempted to press my panic button, but as much as I might have wanted this, there seemed to be something preventing me. I could only surmise it was the presence of the goddess in the shell who was the cause.

The sound in the shell was now becoming quite intolerable to my ears. At first it had intrigued me by the whole range of its frequencies, and like the French engineer, Professor Gavaud, who had built a giant police whistle powered by an air compressor, the testing of which had caused his poor technician to die from the scrambling of his organs, I wondered if I too would die now from this increasing and maddening sound.

But it was the rest of my body that was now beginning to ache. For I was beginning to feel as if all of my body's acupuncture points were being vibrated to a fever pitch, as if my body and mind were being 'fined tuned', so to speak, as if to make contact with whatever was in the shell.

Then as if that was not enough for me to wish I had never been born, or consented to being the professor's proxy, I saw materializing before my eyes what appeared to be a rainbow of colors throughout the shell. It was as if a balanced oscillation of harmonic relationships were taking place between our world and the one I knew I was now entering. What this all meant at the moment I could only guess. And the more I thought about what was occurring, the more I began to wonder if my mind were playing tricks with me. Would I begin to see those gremlins next? I wondered, which were said to plague jet pilots as they flew their planes at record-breaking heights?

Yet no sooner had that thought come into my head, than I saw before my eyes what appeared to be the supposed feminine archetype of my unconscious mind! The very thing Elaine had warned me about! And whatever else I was now seeing, most definitely a woman, was making herself seen in just about every which way I turned. She seemed to be everywhere in the shell, as if she were also inside the cockpit of my vehicle at the same time. The beauty of this feminine phantom spirit was already more than I could stand - she was that wondrous to behold. And it was then I undoubtedly knew I was in the glorious presence of this goddess of the ages.

I knew I now had only a few minutes of fuel remaining. Would I be able to see this bewitching goddess when the vehicle came to a sudden stop? I was sadly wondering. Or would I now suddenly cross over into this other dimension before my gas tank ran out of its remaining fuel?

It was at this moment of my perplexity that I suddenly heard the concerned voices of both the professor and Elaine. Yet because of this vision of magnificence before me, I was at a loss of words to answer. This goddess of antiquity, coming from out of what could only be a heavenly region above, was dumbfounding me every moment by moment into a state of helpless bewilderment. And I could not help but be reminded of the words of Melville's 'Ahab', who when standing upon the deck of the Pequod during a thunderstorm, had shouted to an outbreak of Saint Elmo's fire:

There is some unsuffering thing
beyond thee, thou clear spirit,
to whom all thy eternity is but
time, all thy creativeness mechanical.
Through thee, thy flaming self, my
scorched eyes do dimly see it.

Now when I looked upon this goddess of beauty, I noticed she was of a radiant luminescent form, having an aura of brilliance surrounding her, clad in a garment of white with a mantle of the same. That her hands were pressing together in an attitude of prayer, her head looking up and giving off rays of light creating the appearance of a crown. But of the specter of her face, I saw its phenomenal splendor only half of the time, as I was moving along the strip one moment right side up, and the next moment upside down.

Suddenly and without warning I began to feel a terrible prickling pain coming from my waist and going rapidly downward, then the beginnings of spasmodic jerking in my muscles and solar plexus, immediately followed by feelings of electric shocks running up and down the length of my spine. And as a low frequency of current began making its way through what felt like every nerve in my body, I noticed I was beginning to break through some kind of an electrical barrier - my entire cellular structure reacting now like a charged-up battery. Why, even the vehicle was beginning to give off a variety of multicolored sparks, as if in direct response to what was occurring.

I could not help but hear myself moaning and groaning now as I raced around the strip for what seemed the twentieth time to my aching body and mind. When would my gas tank run out of its fucking fuel? I desperately wondered.

Then, as if my pain were now acting like a signal to the goddess to come to my aid, I saw the vehicle become suddenly enveloped with a white cloud of luminescence. And it was then, I was dumbfounded to observe, the vehicle was no longer riding upon the metal tracks of the strip, but was instead floating in midair! I knew this for a fact because I could not see the Moebius strip any longer - not so much as a damn foot of it! In fact, I could not see more than a few feet in either direction. Did I dare lift up the glass canopy of the vehicle now? I was asking myself. For I knew most definitely the vehicle had now come to a complete stop and the engine was no longer running.

Where the hell was I for Christ's sake? But then I remembered I had a radio! And immediately I pressed down on the mike's button and screamed, "Elaine! Professor Moebius! Can you hear me?" But not receiving a reply, I could only try again and again. But again, it was to no avail, I could not get them to answer.

I knew I had to get out of the vehicle now whether I liked it or not. It was the only thing to do because of my pain. And besides, maybe I was in the shell after all. But if I was in some other dimension now, as I knew Professor Moebius prayed, then I knew I was in for the time of my life. And my only hope of getting out of the place was that beautiful apparition of a goddess I had most certainly seen, who now, when I looked all around me, had mysteriously vanished from my sight.

I then realized wherever I was, was being illuminated, not by the string of electrical light bulbs that had strangely disappeared, but by some kind of natural and all encompassing light. As I slowly lifted the glass canopy up and over my head, I could instantly tell I was definitely not in the shell anymore, for the atmosphere now seemed to be a lot thinner and the temperature felt like fifty degrees. There was also a mild tingling of static electricity all about, and the entire area I was in was surrounded by some kind of a fog - a fog I just realized was also giving off the odor of roses as if I had just walked into a flower shop, or had a seat atop a float in the 'Tournament of Roses Parade'.

Glancing all around me now, I noticed I could only see about a hundred feet into the distance, no matter which way I turned, which in itself told me I could not possibly be in the shell. I knew this because the viewing area I was now surveying was too large - the entire area being but a flat and foggy plain. When I looked up toward what appeared to be a sky, I could only make out a single white cloud about a hundred feet above - the cloud itself being no more than a mere sixty feet in its pillar shape of a length.

It was at this moment, however, in my utter helplessness, that I decided to get down on my knees and pray.

Chapter Six

An hour passed, it seemed, and I had not ventured more than a few yards from the vehicle - my ego seemingly caught between the conflict of my emotions and my abject fear. In that space of time I had tried my radio many times, but again, to no avail, I could not get it to function. It seemed wherever I was would not allow radio waves to be sent or received. I knew all this time, the professor and Elaine would be in an alarmed state knowing the vehicle and I had vanished, leaving them wondering if they should come to my aid. I prayed they would not - not my lovely Elaine. I knew I was trapped in some inexplicable void where there was little hope of escaping - the vehicle I had arrived in of no use to me now without the gas and metal strip on which it had raced.

I was still at a loss to explain how I and the vehicle had been teleported to this strange and sorry place - it was just too impossible to conceive. Even though I had been told all the theories about this possibility by Professor Moebius, I still never for one minute believed I would ever end up in any other domain than the shell after my ride. Nevertheless, here I was. And without food or water on top of everything, which had been too ridiculous to bring. Yet what did it matter anyway, I would probably die in this crazy flat plain with or without any food or water.

However, there seemed to be something else bothering me at the moment; something I had done when I had first arrived and exited from the cockpit of my vehicle. And when I glanced at my watch again, I suddenly realized what that perplexing something was: It was still 11:10 in the morning!

Checking my watch to see if it was ticking, I noticed the second hand was still moving. But instead of it moving on to the next second of time, I was amazed to find it moving only in a back and forth motion like a perpetual seesaw, as if the second hand was undecided as to which way to go. Did time in this spot only oscillate back and forth in an eternal now of endless perpetuity? I could not help but wonder. If this was the case, then would I too just eternally oscillate back and forth in this timeless place like the beats of my frightened heart? The very thought of being trapped in this limbo without dying was worse than all my fears of hell. At least in hell there were other humans around and walking; at least in hell one knew one was there for a reason. But what ungodly reason had caused me to be sentenced to this sorry place? Had I perhaps angered the Lord Almighty for tempering with His universe? Surely God wasn't that revengeful that He would resort to banishing me here for wanting to experiment with His celestial laws.

But was it really God I had offended? I shuddered at the thought it was the devil I had angered. If this was actually the location of his hell, then I was most certainly the first living human to have entered. On the other hand, was I really alive?

I immediately began to check my wrist for a pulse, thanking God I could still feel one. But perhaps I was just dreaming? I then reasoned, having only fallen asleep during my ride. But if I was still dreaming, then this was the most realistic dream I had ever had.

As all these worried thoughts ran through my head one after the other, I half expected to come face to face with either the devil, or the fictional movie character of Mr. Jordan. The fact I had already seen the movie - *Here Comes Mr. Jordan* - at least a dozen times and loving Claude Rains' unforgettable portrayal of him, probably had something to do with my sudden expectation of meeting him now.

Then almost at once, passages of Whitman's poem *Leaves of Grass* began to come into my head - all of them seemingly having a direct bearing on my predicament, death, and my immortal soul:

I wander all night in my vision,
Stepping with light feet … swiftly and noiselessly stopping,
Bending with open eyes over shut eyes of sleepers;
Wandering and confused….lost to myself ….illassorted….contradictory,
Pausing and gazing and bending and stopping,
Do you suspect death? If I were to suspect death I should die now.
Do you think I could walk pleasantly and well suited toward annihilation?
O my soul! If I realize you I have satisfaction,
Animals and vegetables! If I realize you I have satisfaction.
Laws of the earth and air! If I have you I have satisfaction.
I cannot define my satisfaction….yet it is so,
I cannot define my life….yet it is so.
I swear I see now that every thing has an eternal soul!
The trees have, rooted in the ground…. the weeds of the sea have….the animals.
I swear I think there is nothing but immortality!
That the exquisite scheme is for it, and the nebulous float is for it, and the cohering is for it,
And all preparation is for it….and identity is for it….and life and death are for it.
To think of time….to think through the retrospection,
To think of today….and the ages contained henceforward.
Have you guessed you yourself would not continue?
Have you dreaded those earth-beetles?
Have you feared the future would be nothing to you?
Is today nothing? Is the beginningless past nothing?

If the future is nothing they are just as surely nothing.
And as to life; I reckon you are the leavings of many deaths,
No doubt I have died myself ten thousand times before.
Eternity lies in bottomless reservoirs…
Its buckets are rising forever and ever,
They pour and they pour and they exhale away.
Now Lucifer is not dead….or if he was I am his sorrowful terrible heir;
I have been wronged….I am oppressed….I hate him that oppresses me.
I will either destroy him, or he shall release me.
Has anyone supposed it lucky to be born?
I hasten to inform him or her it is just as lucky to die, and I know it.
Is the house shut? Is the master away?
Nevertheless be ready….be not weary of watching,
He will soon return….his messengers come soon.

I suddenly began to think of Mr. Jordan again - that last sentence of Whitman's poem I had remembered, probably being the cause. But would the master of this place be as kindhearted and consoling? And where had Walt Whitman come up with these strange passages that had been running through my frightened mind? Had he at one time visited this very place in his poetic reveries perhaps? The spirit that wandered in thought, or in a dream, was known to enter strange realms, it was said.

Yet why weren't there others trapped in this place as well? Did I dare answer that perplexing question now by venturing away from the vehicle I was still too afraid of departing? Because I knew I would only become lost in this damp and steamed up sauna bath, which was reminding me of some foggy day in the city of London.

Then looking down at my feet, which I could hardly see because of the obscuring fog one foot off the ground, I realized I had not as yet bothered to check out the ground upon which I was standing; a ground which now felt as though it were composed of some sort of metal as I bent down and touched it. As I blew away the fog still hiding my view, I was amazed to discover the ground had a bright and silvery color. However, what seemed really odd about this silvery ground was when I began to press down on it, it immediately reacted by giving off blue and fiery sparks, as if the spot I had touched were charged and electrical.

Then, as I started to stand up, I suddenly heard off in the distance what I recognized to be the faint echoing of a voice. And it was of a man with an English accent no less! Now I was really beginning to believe I was in the city of London - the fog everywhere only adding to my belief. So maybe I was now in the 'Underground' railway - the city subway system of London - the silvery ground I was standing upon only the metal platform of some such station: hence the blue fiery sparks under my feet coming from the subway's 'third rail'. But no! That was impossible! For where were the metal railway tracks? I had not noticed any when I had felt the floor, and I had covered a large area with my spread-opened hands. Besides, what was a pillar-shaped cloud doing over my head in London's Underground in the first place?

Quickly, I began calling out into the direction of the voice I was still faintly hearing, "Hello out there! Can you hear me? Is anyone out there?"

As I prayed for an answer to my shouts, I wondered if this man's voice was the voice of perhaps Mr. Claude Rains himself? Had he become after his death, I theorized, the very Mr. Jordan he had portrayed in his film? Had he become an angel for real, in charge of a limbo sort of station for people who had just died? Or was this voice I was still hearing off in the distance I knew belonged to a man, perhaps only a vocal archetype of my unconscious mind? A voice being projected into this void to help to console me?

Suddenly, I heard the man's voice answering back my still echoing shouts with the words: "Yes, I can hear you, old chap! But where in God's name are we?"

With a sigh of relief, I instantly answered, "I have no idea! But keep talking!"

"Yes! That's a good idea. But what do I say to you, Yank?"

"Count numbers for now," I replied. "It's a little early for a conversation, I'm afraid."

Then as I heard my English friend give out a chuckle and count, I carefully began making my way to his merry voice.

As I approached to within ten feet of his happy face, I received what then amounted to the shock of my life. Standing in front of my dumbfounded eyes and wearing an overcoat and hat was none other than the long-lost and presumed dead English actor Mr. Leslie Howard, looking quite young and alive!

"My God!" I said, my mouth half open in awe. "Aren't you Leslie Howard? No! It can't be you; you're supposed to be dead! And besides, you look too young to be him," I said, looking him over carefully now.

"Well why shouldn't I look young and alive, old chap? I'm only fifty years old. But where the bloody hell are we Yank? I was expecting to land in the water when I jumped out of my plane the Jerries just downed. I didn't notice you were aboard though? So from where in the devil, old chap, did you just come?"

While Mr. Howard had been talking, I suddenly realized he was trailing a parachute behind his tall skinny frame.

"You'd never believe me," I said. "But I was hoping you'd be able to tell me where we are? All the same, I think I already know the answer to that question, I'm afraid."

"Well fill me in, old chum, before I begin reciting

Shakespeare's Hamlet, because at this opportune and propitious moment, I can't resist saying the 'To be, or not to be' soliloquy."

"To be or not to be, is not the question. The question is locality, Mr. Howard. Because for starters, I think I should inform you, that about forty-five minutes ago, I was in the State of Texas in the year 1980!"

From the look on Mr. Howard's face, which at the moment was more than a bit amused, for he was smiling, I knew he had not for the life of me taken me seriously.

"1980! Oh come now, Yank. You must be joking. I mean just because neither of us know where we bloody are at the moment is no reason to pretend you're in a movie now. I'm the one to be doing that. And from the smell of roses in this foggy place, I wouldn't be surprised if we're both dead and at our own bloody funerals."

"I'm telling you the truth, Mr. Howard. And I only wish this was a movie set we're now in, because then we could yell 'cut' and get the hell out of here."

"You mean you're actually from the future, old chum?"

"I guess so, which would mean I must be in the past. 1943, if I remember?"

"But how in the bloody hell would you know that if you're from the future, by Jove?"

"Because that's when your plane was shot down by the Nazis, I believe."

"Then what am I doing here with you, Yank?"

"That's what we have to find out and fast, if we want to get out of this place alive, Mr. Howard. There's no food or water around here, I'm afraid."

No sooner had I said that, in this foggy 'Twilight Zone' of a place, when we both heard in the distance what we knew to be the crying sobs of a woman's voice, followed a few seconds later by what we recognized as the same voice speaking a foreign language.

"Looks like someone else is lost in here, Yank - sounds like a young girl speaking in Russian, I believe."

"Yeah! You're right! Let's hope she understands English, otherwise she might be too afraid to come. She might think we're here to harm her or something."

"Yes, you may be right about that, Yank. But I happen to know a little Russian myself. You see I once played a council general stationed in Saint Petersburg during the Russian Revolution in a film of mine called, *British Agent*. So let's see if I can remember what the Russian words for 'over here' are."

Before Mr. Howard could call out those particular words, we heard the young girl's voice yell, "Where are you? Please help me! I am lost! Please tell me where you are!"

After hearing the frightened girl shout out those English words, we both looked at each other, quite relieved, and began shouting, "This way! Over here! We're over here!"

Keeping this up for what seemed a minute, we suddenly saw the girl coming our way through the heavy fog, which was rapidly fading behind her into a fine mist. It was then that I turned to Mr. Howard and said, "Was that girl aboard your plane, Mr. Howard?"

"Why no? I don't believe so. There wasn't anyone aboard resembling her. I wonder who she is? And how she also got lost in here?"

"Well, whoever she is, she sure looks mighty happy we're here. Why, look at that smiling face," I remarked, just as the young girl, who was dressed in a brown parka, black pants and boots, came before us and said with a Russian accent.

"Thank God I have found you! My name is Anastasia, daughter of Tsar Nicholas Romanov the Second, Tsar of all the Russias."

Chapter Seven

After having overcome our complete shock of hearing the girl tell us her famous name, Mr. Howard and I patiently waited for her to continue speaking. But getting only a wide-eyed expression from her, Mr. Howard began saying, "Well well, Your Majesty. Welcome to our little club. But may I ask you how it is you haven't aged all these many years? You should be at least forty years old by now, I would think."

Not wanting to accept time was out of whack in this foggy wasteland of a place I turned to Mr. Howard and said, before the girl had a chance to reply, "Wait a second, Mr. Howard, but I think you should ask her what year it is, because that might explain why she hasn't aged."

"Yes, yes, old man. I think you're right. And if she says what I suspect she's going to say, then I think we're all in a fine mess of trouble, as if we weren't in one fine mess already."

Turning to the girl again, Mr. Howard asked, "Excuse us, Your Majesty, but would you mind telling us what year it is?"

Looking at us as if we were two loonies in need of a psychiatrist, the girl replied, "Why, it is October 17, 1918. But why are you gentlemen asking me this foolish question?"

Turning to me at once, Mr. Howard said softly, "Good God, Yank! I thought you were bloody crackers a minute ago, but after hearing her say that date, well I guess you were telling the truth about what you said. I guess H.G. Wells was right all along about time travel. And if this girl is really who she's claiming to be, and you're actually from the future, then there's no telling who we're going to run into in here."

"I believe it was also H.G. Wells who said, 'There are universes that are farther from us than nebula yet all the while closer to us than our hands and feet.' But I should have told you to look at your wristwatch, Mr. Howard, to convince you time as we know it is nonexistent in here."

As I observed Mr. Howard take my hint and look at his watch, I knew from his startled expression that his watch was behaving like mine. We then heard the young girl say, "Gentlemen, but what is the matter? You look so scared. Is it because you are lost too? Like I am with you?" she sympathetically remarked.

"Yes, Your Majesty," Mr. Howard said, "lost we are also, I'm afraid. But there's no need to worry, because my friend and I are going to get you out of this place. I promise."

Turning to Mr. Howard, I whispered, so the girl wouldn't begin to cry. "Mr. Howard, I think I need to have a long talk with you. I mean you don't seem to realize what's happened to us. This place we're in isn't on any map I know. And I'm afraid we might be trapped here for some time. If time is the correct word to describe time as we know it to be."

"Yes, Yank. I've already gathered that bloody fact, but there's no sense in telling the girl about this. Maybe there's a chance she can help us out of here though. Let me ask her how she ended up in here... But maybe I should ask you that question first. And by the way, do you happen to know who won the war?"

"We won it, Mr. Howard, thank God. And as far as answering your other question is concerned, well that might take some time to explain. So you had better ask the girl how she got lost in here. Then maybe I might be able to figure out what to do afterwards. Because I happen to know how I ended up lost in this steam bath. I happen to be a scientist. And you're correct about H.G. Wells - time travel is indeed possible. The machine that marooned me here is not too far from where we're standing. But I'm afraid it's not going to be of any use in getting us back to our times. It's not really a time machine, you see."

"That's a bloody shock to hear… Mr.…?"

"Oh, my name is Dr. Philip Grieg. But I think you had better ask the girl the circumstances under which she arrived in here. We may be wasting valuable time by this chitchat."

"Yes, Dr. Grieg, you're probably right. We can get acquainted with each other later."

Watching Mr. Howard turn to the girl with what amounted to an intoxicating smile on his face, I heard him gently ask, "Your Majesty, can you be so kind as to tell us how you became lost with us here?"

Looking at Mr. Howard with the saddest brown eyes I had ever seen, Her Majesty, a girl who looked no older than sixteen to me, calmly replied, "Well gentlemen, I was with the two Russian guards that found me and nursed back my wounds after Mama and Papa, my brother and sisters, were shot to death by the Bolshevik revolutionaries four months ago. And while the guards and I were escaping to the Rumanian border today, we were separated by a dense fog. When the fog cleared I saw a white cloud come down from the sky and envelop me and the next thing I knew I smelled roses and heard your voices in the distance."

"Did this cloud, Your Majesty, look anything like the one above our heads at the moment?" I asked.

"Why yes! And it was as long and pillar-shaped like that one as well."

Turning to me again, Mr. Howard said with a start, "Dr. Grieg, I think you may be on to something because I seem to remember a story about six clouds, which during the battle of Gallipoli in 1915, were said to have come out of nowhere. Then one of these clouds landed on a hill, where moments later, a whole regiment of British soldiers, comprised of between eight hundred to four thousand men, marched calmly right into it in single file. But the crazy thing is that when this cloud then lifted up to rejoin the others, now traveling against the wind, all these men, from the First Fourth Norfolk, I believe, were nowhere to be found. They had all disappeared right off the face of the earth....and I might add, that all this was observed by other soldiers of a New Zealand division positioned on a ridge overlooking the terrain, proving this actually occurred. Because after the war, the Turks, when the British government inquired of these lost soldiers, replied they had neither captured that regiment, made contact with it, nor even knew it existed."

"But even if all that's true, how do you explain how you arrived here yourself?" I asked.

"Well I believe I may have parachuted right through that very cloud above our heads. Then again, I seem to remember, from the Bible this time, that clouds were often used to transport angels, as well as our Lord Jesus Christ, the prophet Elijah, and the prophet Moses. And then there's that pillar of cloud that seemed to follow the Israelites in the desert, of which the Bible speaks..."

"My God!" I almost yelled. "That would mean that cloud above us might be the same cloud - the 'Cloud of the Eternal'!"

"Excuse me, gentlemen," we suddenly heard Her Majesty say, "but isn't that a man coming out of the fog over there?"

As we all looked in the direction Her Majesty was now pointing, we observed a man of about forty-five heading our way wearing what looked to be a bomber jacket and trailing a parachute behind him. He seemed to be waving his arms above his head as if trying to gain our attention, also like he was thinking we were enemy soldiers and letting us know he was unarmed, or just waving to let us know he was there.

Suddenly, we heard him shout out in the French

language, "Dieu meri J'ai trouve' quelqu'un! Dieu merci! Dieu merci!"

Not understanding what he had said, I turned to Mr. Howard and asked, "What did he just yell? I'm afraid I don't understand French, Mr. Howard."

"He said, 'Thank God I found somebody! Thank God! Thank God!' At least that's what I think he said."

"Well maybe you should yell something back to him in English to see if he knows some."

"Yes. I should have thought of that, Dr. Grieg."

I then heard Mr. Howard yell out to the man, who was about eighty feet away, "Do you understand English, my friend?"

Instantly the three of us heard the man shout, "Yes! Yes I speak English monsieur! But where are we? We should all be in the Atlantic Ocean now because I was just shot down over it by a German Messerschmitt fighter."

"Looks like that makes two of us, Dr. Grieg," Mr. Howard turned to me and said.

"Yes, so it would seem," I answered. "Why don't you ask him that disturbing question now. Because I'm almost certain when you ask him to call out to you the month and the year, that he's going to tell you a different date from the one you and Her Majesty just gave."

"I'm almost too afraid to."

"Then let me...Here he comes now. And from the look of perplexity on his face, I don't think he's going to believe what I have to say anyway. At least not until I show him the dates on my dollar bills, or your passport with the date you entered Lisbon. Unless of course, he happened to have been shot down months after you and already knows about your disappearance."

"But how in the world did you know I took off from Lisbon, Dr. Grieg?"

"Oh, I happen to have read your son's book on

your life. He became an actor himself. And quite a fine one, I might add. He had a series on the BBC where he played Sherlock Holmes, I believe."

"That's quite remarkable, I'm happy to say. Thank you for telling me, Dr. Grieg. I only pray we'll get out of this place alive so I can see him and my wife again, which doesn't look too promising I'm afraid, because I don't believe even the likes of Mr. Sherlock Holmes, who my son portrayed, could figure a way out of this baffling case."

Almost on the verge of tears because of what Mr. Howard had expressed, the thought of not seeing Elaine again entered my mind, as I then said to the French pilot now standing some three feet away, "Welcome my friend. We just arrived here ourselves…Wait a second, but your face looks familiar to me," I said, his name on the tip of my tongue. "No! It can't be! Aren't you that famous writer who wrote *The Little Prince*?" I finished saying, still studying his features, as he then looked at Mr. Howard with a shocked face.

"Aren't you Leslie Howard, monsieur?" he said, more than amazed. "You're supposed to be dead. Didn't your plane vanish a year ago?"

"That depends on what date you tell us is today, my friend. But it would seem you're from the future from the looks of it, Mr. Antoine de Saint-Exupéry, if I'm not mistaken?"

"Yes!" I said. "That's who he resembles. I almost forgot. His plane was lost over the Atlantic in 44. But I'm forgetting about you, Mr. Howard, and Her Majesty. So I guess it's him I'm seeing. That is if he doesn't have a double or a twin brother?"

"No. I have no twin brother monsieur. However, you gentlemen are correct. I am Antoine de Saint-Exupéry. But who are you, my friend?" he asked, looking at me and then at Her Majesty and saying, "And the young mademoiselle, Mr. Howard? That is who you are, is it not, monsieur?"

"Yes, it is, Mr. Saint-Exupéry. And this gentleman is Dr. Grieg. He's an American scientist. And the young lady is none other than Her Royal Majesty, Anastasia Romanov, daughter of the murdered Tsar Nicholas the Second, hard as this is to believe."

"But how can this be? The mademoiselle looks so young, and was said to have been killed with all of her family twenty-five years ago."

"Yes, I know, old chap, but as my American friend just explained, we're inside some kind of a time warp. And it may have something to do with that cloud above. You wouldn't recall if you parachuted through one after being shot down in your plane, would you, Mr. Saint-Exupéry?"

As Mr. Howard had been talking to our new friend, I could not help but notice the Frenchman's nose. It was a nose that was sort of short, not coming down near enough to his upper lip, but instead, leaving a gap that was unusual. As was the fact his nose kind of tilted up abruptly at its tip, almost exactly like the nose of the cartoon character: 'Mickey Mouse'. Then there was his deep black eyes I could not help but notice as well; eyes that seemed to jut out at you in their sockets; eyes that gave him a kind of luminous gaze, which combined with his short neck that seemed to thrust down on his shoulders, gave him a sort of permanently startled look.

I also noticed he was no man to want to make an enemy of either. He was a tall fellow, having a head of balding black hair, big broad shoulders, making him look almost like a heavyweight boxer because of his large and stocky size. I wondered if he ever had to react like a boxer and punch someone out when they had foolishly commented on his ungainly appearance within his hearing range.

I then noticed a puzzled look on Mr. Saint-Exupéry's face, as he then replied to Mr. Howard's question, "Why yes, monsieur, I do remember falling into such a cloud after jumping from my plane. But why do you ask this question? Has the cloud above something to do with us being here, I believe you said?"

"Yes... Dr. Grieg and I happen to think so."

"Wait a second," I said. "I just remembered I once saw a 'Twilight Zone' episode on TV that had to do with a World War One fighter pilot who flew into a cloud during an air battle and ended up in the future. He then had to fly back again into the same cloud to get back to his time and the same battle. I believe that TV episode was based on the experience of Air Marshall Sir Victor Goddard, who became lost flying over Scotland during a heavy storm. He claimed to have seen an abandoned airfield whose location he knew. But instead of it being deserted, as it was at the time, it was busy with mechanics working on yellow painted planes. It was four years later that this field was re-opened as a Royal Air Force flying school.

"The thing that proves Air Marshall Goddard had flown into the future was the color of the planes. It wasn't until 1938 that all the training planes were painted yellow. They had all been silver in color before then... But I'm forgetting that's not how I arrived here myself. I'm forgetting about the shell and that beautiful goddess I saw."

"What shell and goddess, Dr. Grieg? Are you feeling all right? And what's this thing called 'TV' you just mentioned?"

"Oh that's an invention called 'Television'. It's like going to the movies, only you watch the movie show at home on a phosphorescent screen about two feet wide. The picture is transmitted through the air like radio waves... As far as the 'Twilight Zone' show is concerned, well that was just one of the many programs being broadcast to entertain people during the 1950's."

"My God! Monsieur! Is that the time you are from?"

After hearing Mr. Saint-Exupéry ask that question, I saw Anastasia suddenly look at me with what amounted to a wide-eyed expression of awe, exactly equaling the Frenchman's.

"Well, not exactly, Mr. Saint-Exupéry," I said. "I'm from 1980, to be more exact. But that's of little importance at the moment. What's important is finding a way out of here and back to our times, if such a thing is heavenly possible. Then again, maybe it is."

"What do you mean, Dr. Grieg? And you forgot to answer my question about this shell and goddess you mentioned."

"Oh yes. That's a long story, I'm afraid. One that you, Mr. Saint-Exupéry, and Her Majesty are not going to believe."

"Well whatever it is you don't think we're going to believe, certainly isn't going to be any harder to believe than what's already happened to us," I heard Mr. Howard say, as he then turned to Mr. Saint-Exupéry and added, "Oh I forgot, Mr. Saint-Exupéry, but do you happen to know what time it is?"

From that particular question Mr. Howard had just asked, I knew he was wondering if the Frenchman's watch was behaving like the both of ours were. And from the look of amazement on the Frenchman's face as he took a look at the time, we knew his clockwork was just as mixed-up as our own.

"What is wrong with my watch? It seems to be broken, Monsieur Howard," Mr. Saint-Exupéry said, obviously confused, as we all observed his eyes moving rapidly back and forth at probably his watch's second hand.

Suddenly, we all heard in the distance the sound of a trombone blaring. It was a single long held note being played.

"Good grief! But is that the sound of a trumpet I

hear?"

"Oui oui, Monsieur Howard, but how can that be?"

"Perhaps it's the Archangel Gabriel blowing his horn. Because I'm beginning to think we're all dead and in heaven now."

"You may be right, Mr. Howard. And I believe I see a man heading our way. We'd better yell to him. He may not have seen us," I said, trying to act calm.

As we began shouting in the man's direction, "Over here! Over here! We're over here!" we heard the man answer back loudly.

"I see you! I see you! But keep shouting! Keep shouting!"

Seconds later, as we saw him finally walk to within throwing distance, wearing what looked like an army officer's uniform and carrying what I recognized to be a trombone in his hand, and then coming face to face with us, I began experiencing the uncanny feeling of having seen him before. It was when I noticed he was wearing round metal-framed glasses that his name came immediately into my mind - that shinny trombone in his hands certainly being the dead giveaway to his world renowned fame - if the word 'dead' was the correct word to use in the matter. But before I could pronounce my name to introduce myself, he said, "The name's Glenn Miller ... but... I should be in the water right now. The plane I was in just went down over the English Channel because of the weather. The wings had so much ice on them I thought they were going to break off. But where the hell are we? Are we dead or something? I mean, who in God's name are you folks?"

"No, Mr. Miller, we're not dead, thank God," Mr. Howard said. "But it's almost as if we all were, I'm afraid to say, because as my companions can attest, we just arrived in this limbo of the lost ourselves, I myself after having had my plane shot down by the Nazis, the young woman arriving after having walked into a mysterious cloud, like the one above, my friend Dr. Grieg, after having been transported here in his time machine that's now inoperable, believe it or not, and Mr. Saint ..."

"Time Machine!" we heard Mr. Miller shout. "You've got to be kidding? Wait a second... You're Leslie Howard! The movie actor! Aren't you?"

"Yes, old chap, I am. But don't let that worry you, because this young lady here is none other than the Grand Duchess Anastasia. And I wouldn't be at all surprised if the next person to arrive in this mixed-up place was the Queen of England herself... But where's your bloody parachute, old man? You didn't just jump out of your plane without one? Did you?"

"No! Thank God! I took it off somewhere in that fog back there. But I wonder if the pilot and colonel who were on my plane jumped out before it crashed. And if they did, I wonder if they're lost in this fog trying to find me now."

"I wouldn't be at all surprised, Mr. Miller," Mr. Howard said. "All the same, you had better start alerting them by blowing your trombone again. Wasn't that why you were blowing it in the first place? And that reminds me, there were passengers on my plane as well, so you had better blow that trombone in every direction, and not just in the one you just came."

Chapter Eight

As Mr. Miller began blowing his trombone, I could not help but remember how my father and I had loved to listen to the music of this famous man and his band. And now, here he was alive before me as if he had not aged a day since he disappeared. What the hell was going on in this screwy mixed-up place God only knew? And why were all these people here with me now? People who had mysteriously vanished? I wondered. Would there be other lost souls showing up here to join the rest of us? I shuddered at the thought. Who wouldn't shudder wondering if 'Jack the Ripper' might be the next to appear? Because there was no telling the kind of people that had been transported to this place via that pillar of cloud hovering above like a 'Goodyear Blimp' in the sky.

Was I right about the cloud above being the 'Cloud of the Eternal'? I wondered. Or was it what I could not bring myself to believe: an alien cigar-shaped UFO from another planet or dimension? I knew from my reading about them that UFO's were capable of gathering water vapor and holding it in a mist about themselves, thus giving them the appearance of a cloud, the metal silvery floor emitting sparks under my feet perhaps the inside surface of some spacecraft. I did not want to admit to myself but a very large sphere would be almost as flat as a level plain if we were standing in a very small portion of it. And the blue fiery sparks would then be the natural result of the electrical phenomena in the air, due to the sphere's electromagnetic propulsion, I presumed.

Then again, what if the Lord God Almighty had his own fleet of UFO spacecraft that looked like long white pillar-shaped clouds, piloted by of all things, that angelic goddess I had seen? Or was it only God Himself who drove flying saucers? I was thinking, as a book on the subject had tried to imply, the silvery floor we were standing upon only a 'cloud with a silver lining' of some sort, as they say? - The blue sparks perhaps only the first telling sign of a static buildup of lightening the cloud was preparing to vent.

All this speculation on my part was just wild theory, and I dared not tell my newly found friends of these conjectures of mine, unless they think I was insane. I had to remember: people of their time, as well as my own, considered people who talked about beings from outer space as being out of their fucking minds.

Suddenly, I heard in the distance again, through the thick heavy fog and from many directions, the shouting voices of many people.

Had Glenn Miller's loud and piercing trombone alerted his lost pilot and colonel, Mr. Howard's fellow passengers and crew, other people lost in the fog from yet other times, or everybody combined? Because from the continued shouting all around, it certainly seemed that way. And when less than a minute later the others and I began to vaguely make out the fog enshrouded figures of many people coming toward us from every direction, there seemed little doubt Mr. Miller's trombone had done its job of summoning to us these newly lost souls.

As I watched them all slowly converge upon us as though they were the Sioux Indians and we were 'Custer's Last Stand', I immediately realized I recognized some of their faces; faces of people I knew I had seen in newspapers, movie reels, and on TV; people who had been considered long dead and mysteriously missing like the people already here; and people who I did not want to meet or strike up a friendly conversation with - at least not with some of them whom I was about to greet with a cold and clammy hand.

Suddenly, I noticed a lone woman among the group of people heading my way dressed in aviator's garb: a tan leather jacket, brown slacks, a brown, gray, and blue plaid shirt, and a brown linen scarf wrapped around her neck. She was also a woman who I knew to be an American and had seen somewhere before. Maybe this was because of her short dirty blond hair and beautiful gray eyes, which combined with her skinny tall frame, were making her quite striking to behold. As I searched my mind for the name that went with her unforgettable face, I realized who it was I now knew her to be. I noticed she had an air of authority about her that could not but remind me of her fame.

Then almost instantly the names of some of the other people coming to a stop around my companions and I began to come into my mind and I knew there was going to be trouble very soon. I knew this because some of these people were carrying guns on their person, like the man in the Nazi officer's uniform, and the man dressed in the uniform of an officer of the U.S. Army Air Corps. If that wasn't enough reason for me to break out into a sweat, then what of the other four men, one of whom was wearing a white straw hat, a white silk shirt and gray flannel pants and also carrying a gun; the second man wearing a tan three piece suit and Panama hat, and the third and fourth men dressed as if they had just come back from the North or South Poles.

Yet it wasn't their manner of dress that was bothering me at the moment, it was instead the way they were all looking at each other; a look I knew was anything but love or admiration. Then I happened to notice another man dressed as a monk, he was wearing a long brown habit and adorned a long brown beard. And yet another man standing next to him wearing glasses and dressed in only a white polo shirt and shorts and looking no older than twenty-three. He looked to be more frightened than the other men around him, as I then noticed still another man directly behind him. He was carrying a brown beat-up attaché case and was dressed in a light blue suit, the pockets of which looked completely stuffed.

As I turned from him to stare into the beautiful gray eyes of the famous woman whose name I now recalled, I heard her say in almost a half-conscious daze, "Where am I? Who are you? What's happened? Where's Freddie?"

As I immediately stepped forward to steady her, knowing she was about to pass out and fall, I loudly exclaimed, "Amelia Earhart! You're Amelia Earhart!"

Yet no sooner had I spoken her name, than I heard Mr. Howard speak out with surprise, "My God! This man over here is the Nazi, Martin Bormann! And that man next to him is the lost Norwegian explorer, Roald Amundsen!"

As Mr. Howard finished speaking, I suddenly heard a man dressed in a blue suit and who I had not counted among the men I had seen, shout out to the American pilot six feet away, "And you're Joseph Kennedy Jr.! What the hell is going on here? You were killed in 1944! And so was that woman there! And damn it, how the hell can that guy be Martin Bormann? What the devil is this? Am I dead or what? This is impossible! This can't be real!"

It was then that I knew who had just spoken those angry words. It was a man I did not want for an enemy, even though he was a fellow American, because he was none other than Mr. Jimmy Hoffa himself!

Suddenly the Nazi officer said, with a heavy but clear German accent, "Pardon me, gentlemen, but you are in the presence of Martin Bormann, the present Reich Fuhrer of Germany and the Nazi Socialist Party, and I have no wish to play name games with you all. But perhaps I should first ask Herr Howard, who I know knows who I am ...and there's no need to deny who you are, for I would have recognized your English dog of a face anywhere, just what he is doing here alive with Herr Saint-Exupéry and Fraulein Earhart? As well as with some of the other people here I know to be equally as dead?"

"That will take some explaining, my dear schweinhund fuhrer," we all heard Mr. Howard say. "But first I will take that pistol of yours. That is if you want to get out of this place alive. And that goes for the rest of you gentlemen I noticed have firearms. We can't have any of you start shooting each other in here like it's World War Three. Now can we, old chaps? We have to protect Her Royal Majesty, the Grand Duchess Anastasia, from any harm you might accidentally cause to her, as well as to the lovely Miss Earhart."

As I watched, the man dressed as a monk suddenly turn to Anastasia with a complete look of surprise, as the men with the guns were handing them reluctantly over to Mr. Howard: five guns in all, we all heard the man in the tan Panama hat and three-piece suit, who looked to be forty, six feet tall and slim, say in an alarming tone, "What do you mean World War Three? Am I to understand there was a World War Two already?"

After which Mr. Howard replied, "I take it you're from before World War Two, Mr....?"

"Oh, my name is Judge Joseph Force Crater, judge of the New York State Supreme Court. But wait a darn second that man standing over there looks familiar to me. Please ask him his name, if you will?"

The judge, I observed, was pointing at the man dressed as a monk, who was now looking back at him with a wide-eyed expression, for it appeared he realized he did not fit in with this crowd of people, what with his strange manner of dress.

It was then that everyone heard Anastasia say, while pointing at the man, "He is Tsar Alexander the

First of all the Russias, who defeated Napoleon, then disappeared in 1825, and my great-great-granduncle, the brother of Tsar Nicholas the First, who was my great-great-grandfather!"

Immediately on hearing Anastasia finish speaking, we all saw the man dressed as a monk, who now looked quite surprised, say to her in English in a warm and loving way, "Bless you my child, for only God could have performed this miracle of us being here then revealing to you my name."

"But how can this be?" we all heard the man dressed in the white polo shirt, shorts, and glasses say, as he went on to add, "How can all you people be alive? People I know for a fact are dead?"

"But who are you, monsieur?" Mr. Saint Exupéry asked.

"I'm Michael Clark Rockefeller, the son of Nelson A. Rockefeller, governor of the State of New York. And twenty minutes ago I was in Dutch New Guinea being chased by about a hundred bloodthirsty headhunters. So how in the hell can I be here with you people? Not that I'm complaining about being just saved from a certain death. But if I'm not dead and in heaven now, then just where the devil am I?"

"Yes! Just how in the world did we all end up here, because believe it or not, I was just climbing Mount Everest thirty minutes ago."

"And what is your name, my dear sir?" Mr. Howard asked one of the two men dressed in arctic clothing who had just spoken.

"It's George Leigh Mallory, Mr. Howard. But how is it possible we're here? Like Mr. Rockefeller said, or are we all dead, as I'm sure everyone here wants to know?"

"No, old chap, you're quite alive, as all the rest of us are. But before we all go mad trying to figure out where we are and why, I think we had better introduce ourselves first, those of us that is, who haven't done so.

Then maybe Dr. Grieg and I can try and explain what's happened. My friend Dr. Grieg is an American scientist, who it seems, was the first to arrive in this bloody Turkish bath of a place."

Immediately I saw three men come forward who had not introduced themselves. One was the man in the light blue suit carrying the beat-up attaché case, the second man, a very distinguished person, I had not noticed before, dressed in a blue pinstriped suit, and the third man the American officer of the Army Air Corps.

"I'm D.B. Cooper," we heard the man with the attaché case say.

Followed by the distinguished man saying, "And I'm Raoul Wallenberg, a Swedish diplomat attached to the Swedish embassy in Nazi occupied Hungary."

"And I'm Joseph P. Kennedy Jr., Mr. Howard... Hey! Is that guy with the trombone in his hand Glenn Miller?" we heard the American officer then ask.

"Yeah! That's who he is! I knew I heard the sound of a trombone in this fog back there," we heard the man I knew to be Jimmy Hoffa say.

Mr. Miller then answered, "Yes, I'm Glenn Miller. I guess the pilot and colonel who were on my plane didn't survive, nor the passengers on your plane Mr. Howard."

"Yes. It looks that way, mate. Perhaps we should consider them lucky they weren't trapped in here too. Then again, at least we're alive, even if we don't know where we bloody are."

"I thought you were going to explain to us that mystery, Herr Howard? Or did you have us give up our pistols so you could become the Fuhrer in here?"

"I didn't need anyone's gun if it had been my intention to become the leader of these people, Herr Bormann. But if anyone should be the leader in this place, it certainly isn't you. You Nazi bastard! So I'd be very careful in what you say to us, because I don't happen to see anyone here who is a German, or Nazi sympathizer. And you're lucky no one has tried to strangle the bloody life out of you yet. And you needn't worry about me shooting your bleeding ass because I'll be turning you over to the British or American government the minute we get out of here, now that the war is won."

"What do you mean the war is won, Mr. Howard?" I heard Mr. Kennedy ask, followed by Amelia Earhart and Glenn Miller asking the same.

"Well it seems my friend, Dr. Grieg, is from the future, 1980 to be exact. He arrived here in a time machine of sorts, believe it or not. Unfortunately it's not operable anymore. But perhaps I should let him tell you what he thinks has happened to us."

Nodding my head in agreement to Mr. Howard, I took a deep swallow and said, "Well, I hardly know where to begin. But maybe I should first inform all the people here who were fighting in World War Two, or were transported here before the war, that the war ended in 1945, and that we won it - the allies that is. Now that that question is out of the way, I guess you all want to know how and why we're here. Well, I can't give you all a definite answer, but maybe after I've told you my personal theory, then maybe one of you can come up with another one to match. On the other hand, even if we do figure out what has caused us to be marooned here I don't think it's going to be of any help in getting us back to wherever we came from, I'm sorry to say. However, I wouldn't give up all hope at this early stage because there's a possibility help from my time may be arriving soon."

"Just what do you mean by that, Dr. Grieg?" one of the men in artic clothing inquired of me.

"And what is your name sir?" I then asked.

"Roald Amundsen, the Norwegian explorer."

"Oh yes...Mr. Howard pointed you out...Well, Mr. Amundsen, even though my time machine, which it really isn't, by the way, is quite useless in this place without the gas and metal tracks upon which it used to materialize here, the people I left behind in what I believe is our own dimension, may try and rescue me. That is if they realize it can only be done by bringing along with them the necessary gas and tracks on the other vehicles we had constructed. But even if they do bring along these tracks and we succeed in building another strip upon which their vehicles and mine can ride, we would still need to put the strip in an area of high magnetic force. However, I'm afraid that wouldn't be of any help to any of you if we should accomplish this. Unless of course we can make return trips from my time to this place and bring each of you out. Except you would then be in my time, and I don't know if you would find my time to your liking, or if such a thing like this is even possible in the first place...I mean getting back to the year 1980, that is."

"But the rest of us here didn't use a time machine, or whatever it is, to get to this place, Dr. Grieg," the judge was saying to me. "So maybe if we can figure out the way we got here, we may be able to reverse the thing."

"Yes, Judge Crater that might be true," I said. "But let me ask each of you now if you can remember anything about what you were doing before you noticed you were in this foggy place? I mean did any of you happen to notice a cloud coming down on you? Did you enter one perhaps, because three of us already had this happen to them?"

"Hey yeah, come to think of it, I did notice this cloud on the road when I was high-tailing it out of town to Mexico. I even had to pull over to the side of the road to avoid it. And when I got out of my car this cloud then started making a beeline for me like a bullet with my name on it only at a much slower speed of course. I guess you can't cheat fate. I mean when your number's up, I guess it's up, like they say."

"What do you mean Mr...?" I then asked this mean-looking unshaven man wearing the white straw hat, gray flannel pants, white short sleeved silk shirt, and who had as yet to introduce himself.

"Oh, the name's John Dillinger. Maybe some of you people have already heard of me?"

"Oh yes. The famous gangster," I said, now aware why this man looked so arrogant and hostile. A man who I also knew had been in his time America's most wanted criminal at large during 1933 and 1934. And a man I knew had killed a cop, making Dillinger idolized by most youths because of it, along with his two daring escapes from jail. His first and most famous escape being the one where he had cleverly carved out of a piece of wood, a gun he had afterwards blackened with shoe polish. I also knew he had spent nine years in prison for his first robbery, and had been dishonorably discharged from the navy for being 'absent without leave' years earlier.

"Well anyway," we all heard him continue to say. "That double of mine I had paid to take my place, after I had a plastic surgeon change my looks, didn't do me a bit of good because he got himself shot dead by the Feds in front of that movie theater in Chicago. I was supposed to be there, you see... Then again, maybe he did me some good if the Feds think I'm dead...I guess it was that 'Lady in Red' who was my moll's girlfriend who must have tipped them off. She must have made a deal with them so she could stay in America, the foreign bitch, but what the fucking hell, I'm a dead man in this bullshit of a place anyway. You can't win man. You can't fucking win."

"Hey! You want to watch your mouth, sonny. There are two young ladies here. Or haven't you noticed?"

"Hey! Who you calling sonny, Judge? You're lucky I'm not packing my rod because if I were, I don't think you'd be alive and breathing now."

"Cool it, punk!" everyone heard Mr. Hoffa say, who then sharply added, "Or do you want me to get tough with you? I happen to be an expert in dealing with smart-ass characters. Why just ask Dr. Grieg or Mr. D.B. Cooper about that. They're both from my time and should be acquainted with my life. Because being the former president of the Teamsters Union, I was no man to fool with. And that includes you, punk!"

After hearing Mr. Hoffa, I knew Dillinger would not be expressing himself with any more expletives. I also knew by the look of rage on Dillinger's unshaven face, that he now had it in for Hoffa's blood. And from the way Mr. Hoffa had spoken to him with those hot-tempered words, I instantly understood why he had been elected president of his union: one was just too overwhelmed with his personality to have wanted to cross him.

"Well I guess that explains how you got here, Mr. Dillinger," I said, keeping a watchful eye on his hands, as I added, "and a few other things, I'm afraid. But maybe I should ask you, Mr. Hoffa, about how you ended up in here? Because I would sure like to know how you mysteriously vanished. Your sudden disappearance five years ago is still talked about in 1980. You really did take a powder back then. So what really happened to you, if I may ask?"

"Well, bud," I heard him say with a cynical smile, "I thought I was a dead man forty minutes ago. I mean I was in the process of being taken for a ride, if you get what I mean? But then, suddenly, this white cloud came down from the sky and swallowed me whole. Then the three hoods that had me at gunpoint, paid off to kill me after my prison term so I wouldn't be re-elected, disappeared all of a sudden. And the next thing I knew I was walking alone in here, but how's about asking Mr. Kennedy Jr. his story next. Because something tells me there's more to this place we're in than meets the eye. I mean ain't it kind of extraordinary that I'm here, with of all people, the brother of my most hated enemy?"

Everyone then heard Mr. Kennedy say, as I turned his way, "What do you mean by that, Mr. Hoffa? Is there something I should know about my brothers you're not saying?"

"Perhaps I can answer that question for you, Mr. Kennedy," I said. "But I don't think you're going to like what I say. At least not all of it, I'm afraid."

"Tell the kid, Dr. Grieg! Because if you don't, I sure the hell will!" Mr. Hoffa shouted to me with what amounted to a bitter snarl. And I knew by that snarl exactly what Mr. Hoffa's motive was for wanting to tell Joe Jr. about what had happened to his three younger brothers. No one had to tell me it had been Joe's brother, Robert, who had wanted to destroy Mr. Hoffa when Mr. Hoffa had been before him taking the fifth in Washington D.C., in front of the McClellan committee, of which Robert Kennedy was the chief legal counsel. Mr. Hoffa had taken the fifth because of not wanting to admit to the corrupt goings on in the Teamsters Union, which Robert Kennedy was determined to rid of its known controlling influence by members of organized crime. Not to mention the fact that it had also been Robert Kennedy, I remembered, who had eventually succeeded in getting Mr. Hoffa sentenced to a prison term of eight years for jury tampering, and another eight years for his alleged misuse of the Teamsters Union pension fund, then rumored to be the underworld's bank which they used to finance some of the most famous vacation properties in the world.

And as I remembered it now, Mr. Hoffa had happily beat his long sixteen year sentence and served a total of only five years because of the intervention of the President of the United States, Richard M. Nixon, whom the Teamsters Union had promised to support in the coming election.

So was it any wonder then that Mr. Hoffa now

wanted to take out his pent-up anger on the oldest brother of the Kennedy clan, an anger he now wanted to express by revealing to Joe Jr. all the bloody and gory details of Jack and Robert Kennedy's deaths at the hands of two cowardly assassins.

Yet wasn't that exactly what I myself was about to do to poor Joe Jr.? But I had no available choice in the matter after what Mr. Hoffa had just said to me. Maybe by telling Joe Jr. myself about his brothers' deaths, and not allowing Mr. Hoffa, it would not be as distasteful to him, especially after he learned later from me Mr. Hoffa's real motive in what I was sadly going to say.

So now, as I looked at Mr. Kennedy with my eyes on the verge of tears, knowing he would have most likely become the President of the United States instead of his brother Jack, had not Joe Jr. disappeared into this time trap of oblivion where there seemed to be no avenue of escape, I said in a voice as comforting as I could make, "Mr. Kennedy, I don't know exactly how to tell you what Mr. Hoffa wants me to say to you, in regard to your three brothers, without causing you a great deal of pain. But I guess now is as good a time as any because we may all be dead in a short time anyway if we don't get out of here soon. So I think I should first inform you, that your brother, John, became President of the United States in the year 1961; and that your brother, Robert, was appointed the Attorney General under him after resigning his position as the senator of New York; and that your brother, Ted, is at present, in my time of 1980, senator of Massachusetts."

"Cut the crap, Dr. Grieg!" I suddenly heard Mr. Hoffa yell out to me, then madly add, "And tell the kid how his brothers got it! Or like I said, I'll happily tell him myself. And I promise I won't spare him any of the details either. So be quick about it before I have a change of mind."

"OK, Mr. Hoffa. OK," I replied, trying to calm Mr. Hoffa down, as I turned back to Joe Jr. and said, "I'm sorry to have to inform you, Mr. Kennedy, but your brother, John, was unfortunately assassinated in 1963 during the second year of his presidency by a man some people say was most likely framed for the murder by certain people in the government.... And your brother, Robert, I'm also sorry to say, was himself assassinated sometime later in California while he was campaigning for president by an Arab American who didn't like what Robert had said about giving military aid to Israel. But your brother, Ted, is still living and well and a senator of Massachusetts."

After hearing me explain this sad news to him, I watched Joe Jr. begin to cry, but at the same time say, through his sad and mournful sobs, "Jack! Bobby! Jack! Bobby! No! They can't be dead! No! No! Tell me you've been lying, Dr. Grieg? Tell me it's not true?"

Then suddenly, I heard Mr. Hoffa shout out to him, "It's true kid! They're both dead and buried! And I can't say I'm sorry about it either! Your brother, Robert, was responsible for sending me up the river for five miserable years, the dirty bastard!"

"You piece of shit!" Joe Jr. then madly yelled, as everyone then watched him charge toward Mr. Hoffa and punch him on the jaw, knocking him down to the foggy ground. As everyone watched with amazement Mr. Hoffa then rise to his feet, we saw Joe Jr. hit Mr. Hoffa again on the jaw, knocking him down again, only this time flat on his stomach and out cold on his face.

"What a punch, Monsieur Kennedy," we heard Mr. Saint-Exupéry then exclaim, as everyone began to pat Joe Jr. on his back, Dillinger saying when his turn came, "Hoffa ain't so tough."

After hearing Dillinger say those words, I could not much blame him or the others for wanting to congratulate Joe Jr. Mr. Hoffa had been way out of line in wanting to tell him what I had been forcefully obliged to say. Not to mention his insulting and insensitive remarks to Joe Jr. afterwards.

I was wondering to myself what Mr. Hoffa was going to do when he regained his senses and it was just lucky Mr. Howard had everyone's gun, or Joe Jr. would have probably been dead right now. I knew Mr. Hoffa would have had his hand on his gun, if he had brought one, and shot Joe Jr. the moment he began making a threatening move. Mr. Hoffa must have thought he would be able to punch him out, knowing Joe Jr. had no gun on him when Joe Jr. reacted in the only way he could to those unkindly remarks. I guess Mr. Hoffa had underestimated Joe Jr.'s running speed.

Suddenly pleased with himself after everyone had patted him on the back, with the exception of myself and Herr Bormann, Joe Jr. came toward me with a smile on his face and said, "Sorry about that, Dr. Grieg, but Hoffa had it coming for saying that about Bobby and Jack. But tell me it isn't true what you just said about them?"

As I put my right hand upon Joe Jr.'s shoulder, I said, again with regret, "I'm sorry Mr. Kennedy, but it's true. I know it's hard for you to accept your brothers, Bobby and Jack, are dead, but they're still alive in your time, which should be some sort of relief to you. And I also know it's pretty hard to believe I'm from the future, but that's equally as true. If you have any doubts of this, I can show you some proof. Like for instance, my driver's license, or the dates on the dollar bills in my wallet. Or I could have Mr. Rockefeller, Mr. Cooper, or if I have to, even Mr. Hoffa, show you the dates on their driver's license or dollar bills. These men are all after your time, as well."

"No. That won't be necessary. I believe you, Doc because I can't explain these people being here. Especially Herr Bormann over there...This is really a nightmare, isn't it?"

"Yes, you can say that again, Mr. Kennedy,"

Miss Earhart remarked.

After which I added, "But somehow I know in my gut we're all going to get out of here alive and back to what's reality. At what time and place, I can't for the life of me say."

Then suddenly everyone heard Bormann yell out wildly, "My watch! What is going on here! Why is my watch behaving like Herr Amundsen's?"

It was at that moment I knew the time had come to explain my theory on this oddity. So I began doing so by saying, "There's nothing wrong with anyone's watch, Herr Bormann. It seems time doesn't exist in this void as we know it."

"But if that is true, Dr. Grieg," Anastasia was saying, "then why are we not all frozen or motionless? Time must exist here or we would not be alive right now. Or would we?"

"Agreed, Your Majesty. Time does exist here, but in a much slower way. And to prove my point: why hasn't anyone complained about being hungry or thirsty since arriving here?"

"Why yes! You're right, Dr. Grieg. Just why hasn't anyone complained about that?" Mr. Howard asked me.

"Well," I said, "that's because hunger and thirst are time-related, and there's obviously something wrong with our biological time clocks now. We must be all living in either some eternal now, where there is no past or future, or some kind of perpetual limbo where the present is moving at an infinitesimal rate of speed."

"Wait a second, Dr. Grieg," Judge Crater was saying to me, "but I seem to remember the Bible says 'limbo' is where people go who have committed suicide."

"No one here has done that, Judge," I replied, sort of bewildered.

"Not in a physical sense, Dr. Grieg," Judge Crater continued. "But who's to say everyone here isn't unconsciously harboring the thought. I mean just look at the suicidal danger, you, and the rest of us were placing ourselves in by our exploits. Take me for example. Why, I myself was about to disclose to the world, because of my sworn duty as a judge, how I had been unknowingly involved with the shady dealings of some dishonest men; shady dealings of which these men didn't want revealed for fear of being jailed; and shady dealings I knew they would try to keep hidden from the world by killing the person trying to expose them, that being myself, Dr. Grieg, like I said."

"Yes, Judge Crater," I replied. "I can certainly agree that some, if not all of us were involved in, what to all intents and purposes, were suicidal exploits. But to get back to what I was going to say. I believe it was Albert Einstein who said: 'For believing physicists, the distinction between the past, present, and future are only an illusion, even if a stubborn one.' And come to think of it, it was Einstein who also said time would be nonexistent, or slowed down to almost nothing, if one were traveling at the speed of light."

"But what has that to do with why our watches aren't keeping time, Dr. Grieg? I mean you can't possibly believe this is only a dream we're all experiencing, now can you?"

"Well Mr. Amundsen, what if, and I'm only theorizing mind you, but what if we were all inside some sort of a craft traveling at 186,153 miles a second - the actual speed of light, or close to it? Wouldn't that make our watches behave like they're now doing? Why time, as we know it, had ceased to exist?"

"Yes, Dr. Grieg, it probably would. But you can't be serious about what I know you're cleverly trying to bring everyone to believe."

"And just what would that be, Mr. Rockefeller?" I said, well aware everyone would have thought I was insane if I had dared to say outright what I was now leading myself up to explaining.

"Well aren't you trying to tell us." Mr. Rockefeller went on, "that we're now inside a flying saucer traveling at the speed of light?"

"What's a flying saucer?"

"Spaceships of sorts, Mr. Wallenberg, but now that you've brought it up for me," I said, "that's one possibility to explain our dilemma. I mean just look at the sparks under our feet, if you haven't noticed them already. However, that's not the only possible answer to explain where we are and everything that's occurred."

"Well how about letting us in on this other explanation of yours, Doc?" I heard Mr. Miller then ask.

"Like I said, before I can possibly tell all of you this other theory of mine, I need to know how each of you arrived in here, as well as what you were all doing just previously to that particular moment....So let's continue where we left off," I said, as I turned to Mr. D.B. Cooper and asked, "Mr. Cooper, your story about your disappearance I believe I already know from reading the newspaper accounts. But for the benefit of the rest of us, please add anything that the newspapers couldn't possibly have known about your motive in hijacking that plane you were on...so please tell us what happened to you. And don't lie about the money you received, or anything, Mr. Cooper, because it might be important in getting us home."

"OK, Doc. I guess I have nothing to fear from anyone, not in this place anyway, but do you want me to begin with the hijacking first?"

"No, Mr. Cooper. Start before you got on the plane."

"OK. But shouldn't Mr. Hoffa be awake to hear this too?"

"I guess maybe so. Somebody wake him up. And Mr. Howard," I said, turning to him, "you had better keep a gun pointed at Mr. Hoffa should he get any ideas about hurting Mr. Kennedy."

"Sure thing, old chap that I'll do gladly."

After Mr. Mallory had awakened Mr. Hoffa, we all suddenly heard Dillinger shout out to Hoffa with a smiling face, "What happened big man? You forget to duck?"

I then noticed Mr. Hoffa groggily look Dillinger in the eyes, obviously only half hearing what Dillinger had just yelled. Mr. Hoffa was now just rubbing his jaw where Joe Jr. had slugged him minutes before, and I could only surmise Mr. Hoffa was only semiconscious now, or Dillinger would have probably been out cold on the floor where Mr. Hoffa himself had been.

It was then, however, that I turned back to Mr. Cooper and said, "OK, you can begin relating the events leading up to the hijacking now. Tell us about it please."

"Well, this is kind of embarrassing in front of all these people, Doc. I mean I'm not really a criminal. I did what I did out of sheer stupidity."

"Yeah that's what they all say, mack," Dillinger was saying. "I robbed twenty banks in a year because I was stupid too. I should have become a politician and robbed everybody legally, like Judge Crater over there."

"Shut up, punk!" we heard Mr. Hoffa shout, then add with a growling voice, "Or have you forgotten what I told you? And don't let that lucky punch of Kennedy's fool you. He got one in when I wasn't looking. I'm ready now smart ass."

"He got in two! Fatso! And I'm not scared of you no matter who you are. I eat guys like you for breakfast where I'm from. Just ask Baby Face Nelson, or Pretty Boy Floyd, or Machine Gun Kelly, who I know. So any time you're ready, I'm game. Just say the fucking word."

Knowing Mr. Hoffa and Dillinger were about to exchange blows very soon, I shouted quite sternly, "That'll be enough from you two! If you want to kill each other, then do it after we've figured a way out of here! We may have to depend on each and every person here in case anything unforeseen occurs because we may be only some of the people trapped in here. And we can't be sure they're going to be happy to see us. Furthermore, you both may be responsible for saving the other's life for all we know. So you better start acting more sociable. And that goes for you too, Mr. Kennedy."

"I understand, Dr. Grieg. You won't have any-more trouble from me in regards to Mr. Hoffa."

"Thank you, Mr. Kennedy. And do I have your word also, you two?" I said to Mr. Hoffa and Dillinger, wishing I now had the gun Mr. Howard had aimed at them.

"OK, Doc. I'll lay off Dillinger if he does the same."

"Thank you, Mr. Hoffa," I said as I turned to Dillinger, waiting for his pledge.

"All right, mack. You got my word. For the time being that is," he answered me.

"Fine then, but remember, I have only to ask Mr. Howard to keep the peace with that gun he's holding if it becomes necessary. So, Mr. Cooper," I said to him, "please continue. You should have no more interrup-tions from anyone now."

"OK, Dr. Grieg. But I want to make it clear that the events leading up to why I had to hijack that Northwest airliner had nothing to do with me wanting to make a name for myself. Rather, it had to do with me wanting to start another life. You see, I had broken up with my wife and lost my job through no fault of my own. My wife though, had left me months before I lost my job, but had later come back to me with my son and daughter. It was because she'd taken all of my money out of the bank, refusing to give me some, and claiming she needed it for her and the kids' welfare, that I then asked her for a divorce. When I lost my job as a salesman of industrial chemicals right after this, I decided to just up and disappear. I wanted to leave my wife and kids my G.I. life insurance. In order for me to accomplish this, I knew I was going to have to fake my own death for them to receive the money. The fact that my wife already had all my money and wouldn't give me some really had no influence on my wish to disappear. I just wanted them to have the security this life insurance money would give to them because of my mock death.

"I'd been planning on leaving them for some time anyway, ever since I came across an article in True Magazine. The article was about how one could begin a new life by changing one's name, and then severing all connections with everything having to do with one's life previous to one's disappearance. But I guess you don't want all the sorry facts about this, Doc."

"No, Mr. Cooper, you're wrong. I think it would be better if you gave us a little background about your life first. And I think each of us will have to do the same when our time comes to speak because I'm beginning to suspect we may all have something in common with each other, besides being all trapped in here. When I find what this common link is, then we might have our answer as to why we were all singled out to be in this place. So Mr. Cooper, tell us everything you think had a bearing on your decision to hijack that plane. And if you must go back to your childhood in order to do this, feel free to do so. We've got all the time in this world."

"Wait a minute, Herr Grieg," Bormann said to me suddenly, "do I have to listen to all of you speak about your boring lives, because I am surely not going to tell you about mine. Not that mine is anything but boring, but under the circumstances, I consider myself your prisoner. And according to the articles of war of the Geneva Convention, I do not have to give you anything other than my name, rank, and serial number."

"Shut up, you Nazi bastard!" we all heard Mr. Hoffa shout, then loudly add, "When your turn comes to speak, you had better talk! You're in no position to play games with us! Some of us here already know about your war crimes! So if you want to get out of here alive, you had damn better cooperate! I'm sure none of us men would hesitate in the least about putting a bullet up your fat Nazi ass if you get out of line again! ... Verstehen my fuehrer? You cheap Nazi hood!"

After Mr. Hoffa's last remark, I instantly looked in Dillinger's direction; I knew Dillinger must have taken those last four words very personally. I now feared Dillinger might react in a physical manner by attacking Mr. Hoffa very soon. I knew this by the way Dillinger had already formed his hands into fists, and by the way his face was seething with anger as he looked at Mr. Hoffa with flaring wide eyes filled with a rage.

However, I had little need to worry because if there was one thing I knew Dillinger feared above all else, it was the gun still pointed at him by Mr. Howard.

So turning toward Herr Bormann I said, hoping to settle things down, "Herr Bormann, Mr. Hoffa is quite correct. You had better keep quiet until your turn comes. And when it does, you can either tell us your story, or shut up. If you so choose to hold your tongue, I won't press you to speak. It'll be up to you. You alone will be responsible for your fate. So please no more interruptions from you, or I'll have you gagged. Is that understood? Verstehen?"

"Yes, Herr Grieg. I verstehen," Bormann said to me, now obviously quite exasperated with my warning to him.

Turning back to Mr. Cooper I said, "OK then. Please continue, Mr. Cooper."

"Well Doc, there's not much more I have to say about my life, other than the fact I had read about the disappearance of Judge Crater before the hijacking. That's one of the things that got me to thinking about disappearing myself. I had always believed the judge had vanished voluntarily, until now that is.

"Maybe I should say something about my wartime experiences. I think they may be of some interest to you all. Because there may be a connection between myself and Mr. Kennedy, Miss Earhart, and Mr. Saint-Exupéry."

"Just what might that connection be, Mr. Cooper?" I asked him with interest.

"Well I was a paratrooper in World War Two, in an army airborne division. I saw action at Normandy during the invasion. That's where my paratroop training came in. I used it when I was dropped, together with my unit, over a French town by the name of Saint-Mere-Eglise, which wasn't far from the shore of Utah Beach. I was only a kid of eighteen at that time. I'm forty-five now..."

"Pardon me, Mr. Cooper, but what was the date you entered here? I mean in this place? It might be important," I asked.

"It was, or is, November 24, 1971. The day before Thanksgiving Day, Dr. Grieg. Does that mean anything to you?"

"I'm afraid not as yet, Mr. Cooper. But tell us more about what happened to you after you parachuted into that town. As I remember it, weren't a lot of American soldiers shot dead by the German infantrymen who were there already, while you and the other men in your unit were parachuting down?"

"Yes, Dr. Grieg, they were. Anyway, after I had landed safely, my unit of the 82nd Airborne Division, was suppose to destroy the bridges over the Merderet and Douve rivers near Saint-Mere-Eglize and Carentan, once the causeways that crossed the marshes leading to our landing site were likewise captured or destroyed. Well, to make a long story short, I got wounded shortly thereafter and was sent home with a bullet hole in my left shoulder. It was while I was lying wounded waiting for help to arrive that I witnessed something of a miracle take place - a miracle, that if I hadn't seen it with my own eyes, I would never have believed it..."

"Well just what was this miracle you saw? I'm almost afraid to ask," I said, wondering why Mr. Cooper had paused in his story, as if he were afraid to go on.

"Well, did you ever hear of the World War One story about the angels appearing at the Battle of Mons, Dr Grieg?"

"Why yes. I think I have. Wasn't that the first battle the British fought against the Germans during World War One?" I said, as I observed Bormann strangely give me a smiling grin.

"Yes. That's the battle. And it's claimed by some of the British and German soldiers who fought there that during the British retreat, a British brigade saw ghostly squadrons of cavalry riding alongside their line of march. And that glowing angels, knights, bowmen, saints, and mysterious clouds were also seen, as well as ghostly phantom armies. I think I may have seen something along those lines after having been wounded."

Suddenly I knew I had found the answer I had been looking for. The mere mention of saints, angels, and mysterious clouds was all I had to have heard to make me see the connection Mr. Cooper had with me. As for the connection with all the others here, that I was determined to find out. But did I dare tell any of them about my vision of the goddess at this early stage, without them all thinking I was a loony tune, like I suspected everyone was now thinking Mr. Cooper to be? Well after considering that thought for no more than a few seconds, I knew the answer to that question would be a resounding no - and a resounding yes only after I heard some of the others claim seeing what Mr. Cooper had just disclosed.

So trying to look as though I hadn't been startled

by Mr. Cooper's revelation, I calmly asked, "Just what exactly did you see appear to you while you lay wounded, Mr. Cooper? I'm sure everyone is dying to know?"

"Well, Doc, I think I may have seen an angel appear in front of me after I was shot, but I was losing a lot of blood at the time. So maybe I was seeing things. But I don't really believe that now..."

"Why's that?" everyone heard Mr. Mallory ask.

"I recently talked to a buddy of mine who had been there with me, he'd also been wounded not far from the spot I was shot. He told me only a year ago he'd seen this angel too. He thought at the time he was going nuts, so he never mentioned this to anyone after the war, like I never wanted to either. But when we met at a reunion the 82nd Airborne Division was holding for its World War Two vets, we heard some guy relate he'd seen the same thing. So it was then we both started telling our stories to each other. As it turned out, this angel, who was the exact same angel, had appeared to us all. The reason I know this for a fact was because the description my buddy and this other man gave matched mine to the 'T'! We'd described the same details of her dress and the appearance of her face. And most importantly, we'd described what this angel had done for us while we'd been bleeding to death a mile from Utah Beach."

"And what was that?" I asked, with my heart rate on the rise.

"Well, Doc, believe it or not, she actually stopped the flow of my blood while she was praying at my feet...Now isn't that unbelievable?"

"Yes, it certainly is, Mr. Cooper," I said, quite dumbfounded by his tale.

"Well anyway," Cooper went on, "after we had told each other the details of our angelic encounters, we just looked at each other wide-eyed with amazement for the rest of the day. We seemed too shook up to do anything else. Does any of this mean anything to you, Doc?"

"Yes, it does. All the same, did you ever happen to see this angel again? Like just before you hijacked that plane?"

"No, Doc. Never after that day. So do you want me to tell you the details of the hijacking now? Or does it really matter?"

"I think I've heard what I wanted to hear Mr. Cooper, except how you arrived here. But I believe I know the answer to that question already."

"Oh, you mean the cloud I jumped into after I bailed out from the back of that airliner?"

"Yes. And do you happen to know what that cloud looked like, as far as its shape is concerned?"

"Well, Doc, it looked something like that one above."

"Hey, Cooper! What's in the attaché case and your bulging pockets? It ain't the hijack money is it?" we all suddenly heard Dillinger yell out to Cooper in a threatening way.

"No. There's no money in the case. I got something else inside the thing," we all heard Mr. Cooper calmly reply.

"And what's that, Cooper?" Dillinger yelled rudely again.

"Oh, just the bomb I used to scare the pilot and stewardesses with, Mr. Dillinger."

"A bomb!" I shouted. "Get rid of it now! Before it explodes!"

"Don't worry, Doc," Cooper was saying calmly, "it's not going to explode. It's not even real. It's only three red painted pieces of pipe and a battery. You see? Have a look."

After Cooper had opened his case to show me his fake bomb, I breathed a sigh of relief along with everyone. But then everyone heard Dillinger shout again. "Hey Cooper! Why the hell did you bring a fake bomb with you when you jumped? If it was a fake, then why did you bring it?"

"I wanted to put the money in the case I got from hijacking the plane. That's why."

"Is that the money in those bulging pockets of yours?"

"That's for you to find out Dillinger. But don't even try because I learned how to kill a man with my hands when I was in the war, and I could break your neck with one chop of my hand. Want to find out and see?"

"No thanks, Cooper. Not now. Not while there's a gun pointing my way."

It was then that I turned to Mr. Howard and said, "I think you had better keep your gun aimed in Dillinger's direction from now on. Otherwise he's bound to jump Cooper for that money…and by the way, I think what Cooper saw, and I'm referring to that angel, might very well be the reason why he's here… So let me ask everyone now, 'The Sixty Four Thousand Dollar Question'."

"What the hell is that, old chap?" Mr. Howard asked, as I faced everyone and said:

"Can I have everyone's attention please? And that includes you Mr. Dillinger and you Mr. Cooper. I want to ask everyone a very important question which may help get us home. And it's this: did anyone here have a similar religious experience like Mr. Cooper said happened to him during the war?"

"You mean like seeing a bloody angel, old boy? Is that what you would have us say?"

"Well only if it's true, Mr. Howard. But it doesn't necessarily have to have been an angel. It could have been anything of an unusual nature, something out of the ordinary. Like that cloud up there for instance."

"Well now that you mention it, I do believe I saw a ghost when I was a child, by George!" Mr. Howard said to me almost scared.

"Hey, me too! I saw one also!" Mr. Hoffa said to me, as everyone else responded with similar remarks of seeing something strange in their lives.

I then answered them amazed, "All of you! You mean everyone has had an experience they cannot explain, other than here?"

"It sure looks that way, Dr. Grieg," Judge Crater said, then added, "But we can't all tell our stories at once. So who's the next in line to speak?"

"How about you, Mr. Howard?" I said, turning from the Judge. "You mentioned you saw a ghost in your youth. You want to tell us about it?"

"Well, it was a long time ago, old chum, but I'll try…It happened, if I remember, but how could I forget, when I was taken to this old castle in Scotland…Yes, that's where I saw the ghost. I believe it was a female ghost which appeared to me that day. I think it was right after I had ventured off to investigate the castle's garden. I had started to smell a rose when I saw her all of a sudden. She was dressed, as I recall, in a glowing white gown running from her neck to her feet, and the gown seemed to be reflecting a blue light of some kind. Her hair, I also noticed, was of a golden color and covered by a white kind of mantle which was also giving off short rays of light, by Jove! The effect of these rays made her appear as if she were wearing a crown. I also remember she was looking up toward the sky with her hands together, as if she might be praying. But it was her beautiful face I will never forget. It was so enchantingly resplendent that I was mesmerized to the spot by it. Whoever she was I never found out. Perhaps she was the ghost of 'Mary, Queen of Scots', Dr. Grieg?"

"Hey! That's what I remember her wearing too!" we all heard Mr. Cooper say. "Your description of her, Mr. Howard, is exactly the same as what my friends and I saw…but you said you were about to smell a rose before she appeared, is that correct?"

"Yes, it is. But why do you ask?"

"Well, I just remembered that my two friends and I were laying in a field of roses after getting shot and seeing her…Dr. Grieg," Mr. Cooper said, as he turned to me, "can that fact be of any importance, do you suppose?"

"Well as I remember it," I said, "the smell of roses in any given place where there are no roses to be found is said to indicate the presence of some holy being. And the smell of sulfur, where there is no sulfur around, is said to indicate an evil presence."

"Hey! This place is filled with the smell of roses! And there ain't any around!"

"That's quite true, Mr. Dillinger," I said. "So it would seem something of a holy nature is also occupying this place. And I wouldn't be at all surprised if that angelic being you saw, Mr. Cooper, and the one you saw, Mr. Howard, is causing the smell of roses in here. As to why this being may be around, hiding somewhere, well, that's what we have to discover."

Suddenly, we all heard the man dressed as a monk and who Her Majesty, Anastasia, recognized to be Tsar Alexander the First, remark, "Did you know, Your Majesty, Miss Earhart, and gentlemen, that the body of Saint Teresa of Avila, of Spain, who lies in a church at Alba, has for decades manifested complete incorruptibility accompanied by a perfume of flowers? And the site has witnessed many miracles as well?"

"Yes, I've heard of her, Your Royal Highness," I said. "But it's now been centuries and not decades that her body hasn't decayed, as well as being accompanied by the smell of flowers."

"Your Royal Highness," we all heard Mr. Wallenberg say, "wasn't she the saint that founded the 'Carmelite Order', and who was known to have levitated off the ground many times, like more than seventy other saints of the Catholic Church were said to have done?"

"Yes, Mr. Wallenberg, that was her," His Royal Highness said, then added, "And Dr. Grieg, for example, you should know that roses suffer when there is an even number of them in a vase, and that if you place them in front of a mirror they will die faster, exhausted by contemplation of their beauty? Victims of your negligence, they will think of you with hostility each time they lose a petal prematurely. Crime does not pay, especially if it is the murder of a rose...So is it any wonder then, that the Church of Rome adopted the 'white rose' as the symbol of the Mother of the Savior - hence the 'Rosa Mystica' of the Litanies."

Turning immediately to Dillinger I said, "You see, crime doesn't pay, Mr. Dillinger, especially if it involves the murder of a rose, as His Royal Highness just remarked. Wouldn't you agree my murderous friend?" I finished telling him, as I waited for him to reply to my foolhardy remark, a remark having a purpose to it. I wanted to redirect Dillinger's seething anger away from Mr. Hoffa, who Dillinger had been looking at with a burning hate in his eyes. By my insulting, but true, remark, I knew I would be making an enemy of Dillinger whose hatred I could certainly do without. But at least I had succeeded in getting his attention redirected off of Mr. Hoffa for the moment. As I saw Dillinger look at me now with what I knew to be the same look of hate in his eyes he had been giving to Mr. Hoffa, I knew I had succeeded in my scheme.

Then noticing everyone looking at me with expressions of stunned surprise, I heard Dillinger say to me with his famous newsreel grin, "You shouldn't have said that to me, Doc. I don't like being made fun of, by you, Hoffa, or anyone. But you got the guns on your side, so I ain't gonna budge. Just what did you think you'd achieve by those remarks to me?"

"Only to inform you Mr. Dillinger...Hatred is a frequency. By aiming a point at an enemy with rigorous accuracy, someone with bad intentions can send out malignant waves that are sure to reach their target. They form one of the most accurate weapons, at either long or short range... That you can send out these malignant waves of hate just as certainly and accurately as a speeding bullet. That hatred is a weapon as deadly as a gun, being able to be mentally used at either a long or short range with the same effect. However, love and the power of prayer, thank God, are even more powerful than any earthly weapons....To give you an example: Will you believe me when I tell you that during World War Two a soldier received a burst of submachine fire in his face, at a range of less than fifteen feet, and was unscathed because at that same moment, three hundred miles away, his wife was intensely concentrating on making an armor for him with her love? You will believe me even less if I tell you that when she looked at his photograph, as she did several times every day, she saw that the image of his face had suddenly and inexplicably been riddled with little dots that looked like bullet holes...So you see, Mr. Dillinger, love and the power of prayer are more powerful than the almighty machinegun.

"And if you're wondering why I've gone to the trouble of telling you this, it's because I believe we're all in a place where hate and evil are absent, that is until all of us arrived in here to defile it with our unrighteous and malicious thoughts..."

"If you're trying to tell me we're all in heaven now, then I think you're out of your stinking mind, Doc," Dillinger said to me. "But I liked what you said about that photograph. If only my moll had been looking at my picture like that dame, I could have rested easy robbing them banks. But I've got to admit, crime sure the hell don't pay. Not if it gets you into this kind of mess I mean."

"Well I'm glad you finally figured that out, Mr. Dillinger," I said. "But maybe I should ask you to tell us your story now?"

"Sure, Doc. I'm game. But don't most of you folks know about me already? Hell! I was public enemy number one for almost a year in 1933. And still am, for that matter."

"Yes, I'm sure most of us know about your criminal exploits. Like how you were shot in the chest while wearing a bulletproof vest and then killing the cop responsible. But tell us anything you can think of that happened to you while you were growing up. Anything that was unusual. Like maybe seeing what Mr. Cooper or what Mr. Howard said they saw."

"You mean like seeing an angel? Man, you gotta be kidding. A guy like me seeing one of those things? Never in a million years, but I remember hearing something strange one day when I was a kid of about seventeen. It might have been the voice of my mother, who had died when I was three. Then again, maybe it wasn't her voice I heard."

"What do you mean?" I asked.

"Well, I'd just come out of the pool hall I'd gone to when I saw in the sky what I thought was a plane. Later, when I was nearing my father's farm where I lived, I noticed it again, what I thought was that plane I mean. But now, as I saw it this second time, I knew this thing that looked like a white misty plate, suddenly then start to make a landing about a quarter of a mile away. I immediately began making a beeline for it thinking it was only a damn Zeppelin of some kind. As I got to within an eighth of a mile of the thing, I started to hear this woman's voice in my head..."

"What did the voice say to you?" everyone heard Anastasia ask.

"Well first it called out my name, believe it or not? Then I heard the voice say, 'thou art found wanting', or something, as this misty white plate, which was most likely a cloud, then suddenly took off and disappeared into the sky. I've never told anyone about this before, Doc, and you can see why. Everybody would have called me either a basket case, or just a damn dirty liar."

"I believe you heard and saw what you did, Mr. Dillinger, but what you heard may have only been caused by a feeling of paranoia," I told him, not wanting to believe he had heard the same voice as myself.

"Are you trying to tell me I was just nuts at the time, Doc?"

"No, of course not. But when a person is suffering from paranoia, they frequently have to an acute degree, a power that approaches that of telepathy. A person can imagine they're being persecuted, when in fact, because of a heightened telepathic awareness; they only perceive the unconscious hostility in their parents or loved ones near to them. So did you have, at that time perhaps, a disagreement with your father, Mr. Dillinger?"

"Sure, Doc. I was always having one with him. He wanted me to stay away from the pool hall - he thought I was mixing with the wrong kind of men there. He wanted me to always help him out on the farm, but I just wanted to get as far away from that place as I could get. Doesn't every kid have disagreements with his folks now and again, Doc?"

"Yes, of course. But let me ask you about something you just said. You mentioned that this voice you heard spoke your name. Just what exactly did this voice sound like?" I asked him, wanting to see if he would describe the same voice I had heard in my shell as a child.

"Well, it was like nothing I'd ever heard before. It was like in a dream. Like what you always wanted a dame's voice to sound, but I can't really describe that to you. It's too difficult, and even kind of personal, if you get what I mean?"

"No matter, Mr. Dillinger, you've told me what I wanted to know. That misty plate you saw that might have also been a cloud is sure a good description of a flying saucer. And that's what's bothering me."

"You mean that thing I saw may have been a spaceship, Doc?"

"I'm beginning to think so, Mr. Dillinger. And if it was, then there's a possibility we're all inside of one right now. But it may be a little too early to jump to that conclusion at this time. Still, there's a fifty-fifty chance that that's where we are."

"What's the other possibility you were going to tell us about, Dr. Grieg?" Mr. Miller then asked of me.

"Well none of you are going to like what that other possibility is, I'm afraid, because there's a good chance we're now in another dimension of some kind. You see, time as we know it is not as uncomplicated as most people have come to believe. There may be overlaps between temporal periods, creating on earth unknown and normally inaccessible regions occupying even the same space - a space more complex than we could possibly conceive. There could be magnetic forces in the earth that generate time field anomalies around the entire global system causing ships and planes to disappear," I explained, remembering some of the theories Professor Moebius had told me and Elaine.

"You mean like in the 'Bermuda Triangle', Dr. Grieg?"

"Yes exactly, Mr. Hoffa."

"What's the Bermuda Triangle, Doc?" Dillinger asked me.

"It's a triangular area formed by Bermuda, Puerto Rico and a point near Melbourne, Florida, where ships and planes are said to disappear without a trace. And there may be other areas in the world where the same thing mysteriously occurs. One of these areas, I believe, is called the 'Devil's Sea', located somewhere off the coast of Japan."

"You know, Doc," Mr. Cooper was saying, "I remember reading about a man on a boat who accidentally dropped his knife into the ocean about a hundred miles from Florida. And at the same moment, the man's wife, who was at home, was shocked to then see the same knife suddenly fall from her kitchen ceiling and dig itself into her tabletop about an inch. You don't suppose that guy's knife fell into some kind of inter-dimensional hole, as we all seem to have done, do you Doc?"

"Well there's no other explanation for what happened to that man's knife. Is there?"

"You know, Dr. Grieg," Mr. Rockefeller was saying, "I also read about an incident like that. It happened in 1950, only it involved a man dressed in nineteenth century clothing who looked to be lost in New York, and in his confusion, simply blundered into the traffic in Times Square getting himself killed as a result. But here's the crazy part: this man was then traced back to a missing person case of another man with the same name, and dressed in the same clothes who'd also disappeared, only this other man had disappeared in the year 1879! It seems he'd just gone out for a walk and simply vanished off the face of the earth."

"I disappeared to the world before I disappeared into here, Mr. Rockefeller. So now I'm nobody in two separate worlds."

"Mr. Cooper, if we were nobodies I don't think we would all be here," I said, "because there has to be a good reason why we've all been singled out. And some of us would certainly be dead right now from drowning in the ocean if we hadn't been so miraculously saved. So there must be an intelligence that's responsible for this...But I guess we had better continue telling our stories in order to see if we can find some other clues; clues as to why this intelligence brought us here in the first place. Mr. Hoffa, I remember you mentioned you saw a ghost, like Mr. Howard. Tell us about it please, if you will."

"OK, Dr. Grieg. But I had better not hear anyone start laughing, or there's going to be a guy in here with a bloody nose..."

"We promise not to laugh at you, Mr. Hoffa," Miss Earhart was saying. "But you can't sock all of us in the kissers if we should happen to laugh at your remarks simultaneously, so don't take it too personally. And remember, we're all in this thing together, with the exception of Herr Bormann over there."

"OK, Miss Earhart. You win. I won't punch anyone's lights out. But my story is sort of on the funny side."

"That's a relief," Judge Crater was saying. "That's just what we need to hear at the moment, Mr. Hoffa. So if you can make us laugh by what you're about to say, please do."

"OK, Judge. I only wish you'd been my judge at my trial because I'm sure you would have found me completely innocent...Anyway, I guess you all want to know if I saw an angel? Well, the answer to that question is yes. And I married her soon after."

"Mr. Hoffa, this is no time to make jokes," I said to him as I turned to the judge and nodded, then finished telling Mr. Hoffa, "So if you really saw something like what Mr. Howard or Mr. Cooper saw, please tell us."

"OK, Doc. But you should see my wife. She's really an angel. Anyway, I'll never forget what I'm sure no one is going to believe here. It happened when my daughter and I went to Israel when she was seventeen. We'd just come back from looking over a piece of property I had in mind for a home for children, located on one of the hills of Judea, on a hill by the way, where they say John the Baptist was born. This children's refugee home, I might add, I was going to found with money donated to me at a testimonial dinner in my honor. Well anyway, it was about a few minutes after coming down from that hill that I saw it..."

"Saw what?" I asked.

"I was coming to that, Doc, so hold your horses because I have to fill everyone in on certain details first. You see, after my daughter and I left this famous hill, which had at its summit a strong smell of roses, I happened to glance back at the top and see a glowing white cloud of some kind. I didn't think much of this then, so I didn't mention it to my daughter. However, later that night when I was about to go to bed, I saw the same glowing cloud again, only this time from my hotel room window. And it was then that I had the strangest experience of my life, if not the most frightening. Well, to make a long story short, I began seeing before my eyes, events in my life unfolding, like I was looking at a movie picture screen. I mean as if someone were holding a white five-foot wide screen in front of me, and then running a movie of my life onto it.

"You know, Dr. Grieg, my middle name is 'Riddle', and what happened to me in that room is one riddle I've never figured out. And of course, there's this place here. That's one hell of a riddle too."

"Wait a second, Mr. Hoffa," we all heard Joe Kennedy Jr. now say, "but I remember reading about the very thing happening to a general during the Civil War. I think his name was General McClellan, the Union general in charge of the U.S. Army before General Grant. General McClellan was supposed to have been in an army tent when he had a vision occur before his eyes. He claimed to have seen on a white screen before him, what would have happened if he went ahead with his plans in defending Washington D.C. He said he only saw disaster, so he changed his plans and ended up saving the nation's capital. But I think he mentioned he saw a woman holding up the screen. Or maybe it was George Washington instead. So I'm wondering, Mr. Hoffa, why you didn't see anybody in the room with you?"

"I was coming to that, sonny. You see, Dr.

Grieg," Mr. Hoffa said, turning to me again, "after I'd watched myself going through these episodes in my life on this white screen, like having my scalp laid open six different times by the police, company guards, goons, and strikebreakers because of my union organizing during my first year in the Teamsters Union, I then sadly watched my brother, Billy, get himself shot in the stomach by a company owner's son who had mistaken poor Billy for me during a strike.

"Well it was right after this that I saw this woman dressed in a white glowing gown standing to one side of this movie screen she was all of this time holding. But then, almost at once, she and this screen in my room just up and disappeared. Funny though, but she kind of looked like Marilyn Monroe to me, believe it or not."

"Marilyn Monroe?" I said, realizing at once the being I had seen had possessed her beauty.

"Yeah? Ain't that something? Because even to-day people think I had something to do with her death."

"Why is that?"

"Well Doc, you have to understand that at the time of President Kennedy's administration, rightwing FBI and CIA agents had many reasons to want to bump off Marilyn Monroe..."

"Why was that?" I asked.

"They wanted to embarrass the Kennedy family, as I also wanted to. These agents, you see, were furious with the President and his brother, the Attorney General, over the complete failure of the 'Bay of Pigs' invasion. Robert Kennedy had been against using any air support, you see. And also, because the President, after that disaster, said he was going to limit the power of both the FBI and CIA as a result. Some agents from these two agencies thought, right after this decision of his, that President Kennedy was growing too soft on Cuba, and they were afraid for our country's safety, which was the reason, I think, that they had him killed with the help of the Mafia...You see, Giancana and Roselli, who were two Mafia kingpins, had wanted to recover the Havana vice rackets. It'd been these two guys that the CIA had also used to put a hit on Castro..."

"But how was Marilyn Monroe involved with the Kennedys? I mean, how was she going to embarrass the Kennedys by her death?" I asked, keeping a watchful eye on Joe Jr., knowing he might soon react violently to Mr. Hoffa's disturbing remarks.

"Well, Doc, I thought everyone from my time knew about the affair Marilyn was having with the President and his brother Robert," Mr. Hoffa said, as everyone watched him look toward Joe Jr. with a big smile on his face. "Well anyway," Mr. Hoffa continued, "it's my guess that on the weekend Marilyn was going to give a press conference revealing her involvement with them because of their refusal to see her, Robert Kennedy went to her house to try to get her to change her mind. Since I knew for a fact Marilyn's room had been bugged, I knew the CIA and FBI agents who'd done this, had the very opportunity they'd been looking for. So they force-fed a whole bottle of sleeping pills down Marilyn's throat, or injected her with something similar, to make it look like she'd committed suicide, and then waited for Robert Kennedy to arrive in order to entrap him at the scene. But these agents had other reasons for wanting to kill Marilyn Monroe..."

"Like what?" everyone heard Joe Kennedy Jr. ask, looking to me now like a man who had murder on his mind. And I wondered to myself why he was not challenging Mr. Hoffa's remarks on his brother's alleged adulterous behavior.

"Well for one thing," Mr. Hoffa continued, now directing his attention to me, "she had information she'd written in her diary about the CIA and the Mafia's attempt on Castro's life, enough information to get the President and his brother impeached for probably allowing it. So you see, if Marilyn had lived, she would have done to the Kennedys what I'd always been planning to do: namely, to get even and embarrass them. So I had nothing to gain by her death, Dr. Grieg."

"You know, Mr. Hoffa," I heard Mr. Howard say, "when you made mention of your trip to Israel, you reminded me of this man who'd been on my plane. His name was Wilfrid Israel, and he'd been working against Hitler by rescuing Jews from concentration camps and extermination centers. He'd also been instrumental in helping to set up the Jewish State in Palestine, and had enough on the Nazis to name names, if ever he became a witness at a War Crimes Tribunal. This may have been one of the reasons the Nazis had our plane shot down. Am I correct in this, my dear Herr Bormann?"

"I'm sorry to disappoint you, Herr Howard, but it was only you we wanted to kill. And when I learned you were dead and half Jewish on your father's side, I began rejoicing."

"Shut up, you Nazi bastard! One more word out of your evil mouth and I just might kill you where you stand," we all heard Mr. Howard shout, as everyone then noticed Bormann begin to shake like he had seen a ghost.

"Oh, don't pay him any mind, Mr. Howard. His kind will get their due ... Tell us why the Nazis wanted you so dead?"

"Well, Miss Earhart," we all heard Mr. Howard reply, as he took a glance at frightened Bormann and said, "I suspect it was because I was Britain's most powerful and effective propagandist. You see, I'd ridiculed the German hierarchy in one of the films I'd starred in called, *Pimpernel Smith*. The film was Britain's first World War Two movie and probably the most virulently anti-Nazi film of the year. In it I played a mild-mannered professor secretly working as an undercover leader against the Nazis in Germany.

"And then there was the film, *The 49th Parallel*, I made as well. I had only a small part in that film, in the last quarter of it. In it I played an author who outwitted three Nazi U-boat officers who had abandoned their disabled sub in the Canadian part of the Saint Lawrence River. And while outwitting them, I gave them a long speech of considerable passion, attacking the half-truths and distortions on which the Nazi party had been founded," Mr. Howard finished saying, as everyone then heard Bormann say under his breath:

"Dummkopt."

"I wish I could have seen those two films of yours, Mr. Howard. Because I was something of an undercover agent myself in Budapest, Hungary."

"You know, Mr. Wallenberg, my father was born in Hungary. But what exactly did you do over there?" Mr. Howard asked.

"Oh, I tried to save thousands of Jews from going to the Nazi death camps by giving them false Swedish passports. I could do this because I was pretending to be a diplomat working for the Swedish Embassy in Budapest. And besides, my mother was herself a Jew, so I had a personal motive in doing this," Mr. Wallenberg said, as everyone then heard Bormann give out a laugh.

"Oh I see. All the same, what you did must have taken a lot of guts, Mr. Wallenberg, and I must admire you for it and what you accomplished. But just where did you get the courage to stand up to the Nazis?" Mr. Howard asked, as I watched him then give Bormann the eye as if daring him to speak.

"I have no idea, Mr. Howard. Still, I couldn't just stand by and let these thousands of human beings be sent to a certain death... Your films and your outspokenness against the Nazis likewise took a lot of guts. Didn't you also make a film about the inventor of the British 'Spitfire' airplane, I believe?"

"Why yes, I did. I played Reginald J. Mitchell,

who died of cancer in 1937. He was a self-taught genius who'd never received formal training. Nevertheless, he became chief engineer and designer for Supermarine Aviation Works, the company he worked for. The film in which I played him was called *The First of the Few*."

"Didn't you also play the part of Ashley Wilkes in the movie *Gone with the Wind*, Mr. Howard?

"Yes, I did, Mr. Miller. And come to think of it, I also made a movie that sort of has a lot to do with this foggy place as well."

"Which film was that?" I asked, almost stunned.

"Well it was the first film I made in America. I made it in 1930. It was called *Outward Bound.* And this is what is so unsettling about it: the film was about the possibility of life after death and the migration of souls. The whole story took place aboard an ocean liner adrift in a foggy sea, where all the passengers gradually realized they were dead. Unfortunately, nobody on board that ship made it back to the world of the living...Jesus! Let's hope that doesn't happen to us... And you know, Dr. Grieg, I just realized I seem to have made a lot of movies about people who led lives very much like some of the people here. Take Mr. Dillinger, for example, I starred in a play on Broadway that was later made into a hit movie starring Humphrey Bogart, who was also in the Broadway play with me. The play and movie were called, *The Petrified Forest*. That film was about a gangster by the name of Duke Mantee, played by Bogart, who hides out with his gang at a desert gas station-café in Arizona's Petrified Forest. I played a poet-intellectual wandering across the country who then ends up at this gas station-café before Duke Mantee and his gang arrive. He then later kills me at my request because I wanted to leave my life insurance money to the café's waitress, played by Bette Davis, so she could go to France to study art, believe it or not. The gangster in that movie was patterned after you, Mr. Dillinger, and made Humphrey Bogart a star…

"I also made a film about time travel called, *Berkeley Square*, where I played a man who had received a concussion on the head and was taken back in time to the eighteenth century. Now how's that for coincidences? Oh yes! I forgot to mention a movie I never made. It was called: *The Man Who Lost Himself*…

"So you see, Dr. Grieg, there's something mighty strange with these films I made and a few of the people here, as well as the unusual circumstances. Why even the film I made called *Intermezzo,* about a violinist I portrayed would seem to have a relationship with Mr. Miller."

"Yes, that's truly amazing, Mr. Howard," I said.

Mr. Wallenberg then added, "Mr. Howard, that man who was on your plane, this Mr. Wilfrid Israel, the man you mentioned was rescuing Jews from concentration camps, he also has a connection with me. So might he not have a connection with the others here? What I mean to say is this: is there anyone here, besides you and myself, who might have a connection with us, or with the other passengers who were on the planes some of us were aboard?"

"That's what we've been trying to discover Mr. Wallenberg. That's why Dr. Grieg has been asking all of us these questions, questions that might just shed some light on why we're all here. Maybe Dr. Grieg should ask you now if you saw anything in your life that was out of the ordinary? And I mean like seeing that ghost or angel I know I saw?"

"Yes, Mr. Wallenberg, why don't you tell us your story now. Tell us anything you saw unexplainable in your life that you can think of," I said.

"Well, Dr. Grieg, I did happen to see something of that nature quite recently. But is telling you about things like what Mr. Howard, Mr. Cooper, Mr. Dillinger, and what Mr. Hoffa told us they saw going to really get us out of here because it seems to be leading us nowhere."

"Please, Mr. Wallenberg, let me be the judge of that," I said, "because there's a definite purpose behind all of these questions. And when I tell you my own personal experience with a being I believe I saw, you'll understand immediately why I'm so interested in everyone's stories. So Mr. Wallenberg, please tell us what you witnessed which can't be logically explained, if you would be so kind?"

"Well if you must know, Dr. Grieg, I'll tell you. But first let me ask you a question that's been bothering me. And pardon me if this question of mine then leads you all into asking other questions related to it."

"Why yes of course. What's this question of yours, Mr. Wallenberg?"

"Just this: have you ever heard of the famous Sicilian philosopher, Empedocles, who flourished about 450 BC?"

"Yes, I've heard of him. But why do you ask?"

"Other than being a great philosopher, he was supposed to have secretly thrown himself into the crater of Mount Etna, in order that his mysterious disappearance might be taken as proof of his divine origin. But guess what happened after he did that to himself, Dr. Greig?"

"What?" I asked, wondering what Mr. Wallenberg was up to.

"The volcano threw out his sandals sometime later, and this destroyed the popular belief already established of his divinity."

"It's too bad we all can't be thrown out of here like his sandals were, right Dr. Grieg?"

"That's for sure, Mr. Amundsen," I replied.

"But didn't another famous man also disappear on that mountain in the last century?"

"Yes, in 1889," we all heard Mr. Mallory answer. "But it wasn't on Mount Etna, Mr. Amundsen, it was on Mount Olympus. The chap's name was a Mr. Macmillan, of the famous publishing house. He'd been last seen waving from the mountain's summit, but despite careful searches, no trace of him was ever found - not even his shoes, as poor Empedocles sandals had been."

"A cloud must have got him then. But hey! Didn't this Mr. Macmillan have a son who was an explorer, I seem to recall?"

"Why yes, Miss Earhart! He did indeed! His name was Donald Macmillan," Mr. Amundsen was saying, "and he headed an expedition out to discover a mysterious arctic mountain range that had been sighted by the explorer, Robert Perry. Unfortunately, this mountain range had been nothing more than a mirage in the horizon that he and Perry had been vainly out chasing. But at least he and his party of men survived that expedition."

"You don't suppose, Mr. Amundsen that this place we're in is also just a mirage of some sort, now do you?

"Who's to say, Mr. Kennedy? Who's to say?"

Suddenly everyone heard Mr. Wallenberg say robustly, "That's exactly the very question I was indirectly leading you all up to, Dr. Grieg, the question of this place being only a mirage; a place that could only exist as an illusion in our minds. And not until you, Dr. Grieg, or the others here, give me the evidence to disprove this, will I tell you what I personally witnessed out of the ordinary; something so fantastic that it could only have been a mirage."

"Well I'm afraid that's going to be impossible, Mr. Wallenberg, because no one here can prove any of their theories as yet. All the same, we can all understand your feelings in the matter, but maybe you should first tell everyone how you arrived? Did you happen to see a cloud? Or did you see something else you believe may have been responsible for your circumstances?" I asked him, trying to get Mr. Wallenberg to speak. I suspected it was the presence of Bormann which was making Mr. Wallenberg a might uncooperative, and who could blame him for acting so touchy in Bormann's distasteful company. Bormann represented everything I knew Mr. Wallenberg, the others and myself hated: the Nazis, the Nazi Socialist Party, and its murdering butchers, Herr Bormann included.

As everyone then watched Mr. Wallenberg give Bormann a maddening stare, he then turned toward me and said, "Dr. Grieg, I'm sorry if I've been a bit unrealistic in my request to you, but what I witnessed, I believe could have only been a mirage, was so disquieting to me, I'm more than a little reluctant to discuss it."

"Why is that, Mr. Wallenberg?" I said. "If I may have your permission to ask?"

"Yes of course... Well you see, what I witnessed happened one day when I was in the 'Buda' part of the city of Budapest, on a hill in Buda called *The Hill of Roses*, the hill on which Adolf Eichmann lived in a handsome villa, the very man Hitler had appointed to totally exterminate the Jews.

"I'd been on that hill out of a compelling desire to kill this butcher of humanity. But it was while I was going to his villa, just before I'd reached the top of that hill, devoid of roses, due probably to his evil and coldhearted presence, that I beheld what I still believe could only have been a mirage. And the reason I say and believe this is because I do not believe God would have sent an angel down to earth on this unholy ground knowing Adolf Eichmann resided there...And yes, I do believe in the possible existence of angels, Dr. Grieg, if that was the next question you were going to ask. It's just too distasteful and beyond belief for me to admit to myself that that being I saw was truly an angel for real. It just had to have been a hallucination of my mind. And besides, the whole hill smelled of roses where there weren't any to be found."

"Mr. Wallenberg, the very fact that this hill Eichmann lived on was named *The Hill of Roses* proves that a real sighting of an angel by you more than likely occurred. Because like I remarked earlier, the smell of roses in the absence of them, indicates the presence of something holy, or a deceased loved one in some cases. But tell us what happened the moment after you saw this angel on that hill. It having been only a mirage of an angel, or otherwise, Mr. Wallenberg."

"I saw this angel begin to weep, Dr. Grieg, as if it were crying for what I was then picturing in my mind: images of the dead naked bodies of the many Jews I'd seen floating face down in the freezing Danube River, thrown there in that river by members of the *Arrow Cross* - the Nazi collaborating police of Hungary. This is the reason I believe I was only imagining I was seeing an angel at that time, caused probably by my unconscious mind projecting the image of an angel during those moments of my sorrow and grief. This I believe was in order to comfort and prevent me from doing what I might soon regret, I could then only surmise..."

It was at that moment I was beginning to believe what Elaine had told me about the people of the past seeing a goddess: that they had seen nothing more than a projection of the personification of the feminine attributes existing in their unconscious minds taking on a visible form. But then what had stopped the flow of Mr. Cooper's blood while he was bleeding to death on Utah Beach in Normandy? And caused what Mr. Hoffa and Mr. Howard had seen, and what Dillinger had heard and seen, for that matter? They could not have all been seeing projections from out of their unconscious minds. I was wondering...so did angels really exist?

Then everyone heard Mr. Wallenberg go on to add, "I was also on that hill to kill Eichmann because of the conversation I'd had with him during our meeting the previous day - a conversation that very much contributed to my sad mental state of affairs. You see, I'd told him that the war was almost over, that it was only a matter of weeks before Germany surrendered, and that he should call off his people and leave the Jews alone so he could escape in time from the advancing Russian army. However, the bastard said he was going to stay and do his duty 'til the bitter end. That he wouldn't be satisfied until his 'Final Solution' program took care of exterminating every last Jew left still alive in Hungary and the other Nazi occupied countries.

"You know, Dr. Grieg, I was going to kill Eichmann the moment he existed from his villa, but that being changed my mind. Because I knew, moments later, that if I'd been successful in killing him, I wouldn't thereafter be able to help save anymore Jews from being sent to their deaths. Eichmann would have only been replaced with someone under him just as evil, if not more so, and I would have been shot. Not to mention the reprisals the Nazis would have taken out on my Jewish staff of over four hundred, all of whom I'd saved by issuing to them false Swedish passports. They would have been shot dead in the streets that very day.

"Eichmann had even said to me in that meeting that I wasn't as immune from his authority as I seemed to believe, that my death could be made to appear as an accident at any time he desired. Killing me, he had months before decided, really wouldn't prevent the Swedish embassy from carrying on with its policy of issuing Swedish passports to as many Jews as wanted them. And that my death would only bring to the attention of the world what was really happening to the

Jews, who before I arrived in Hungary, had been going to concentration camps without anyone protesting in the Hungarian government, specifically, the Foreign Minister, Baron Gabor Kemeny. He had only changed his mind about this deportation of Hungarian Jewish citizens because his wife, the Baroness Elisabeth Kemeny, had persuaded him to tell the members of the arm services, as well as the Arrow Cross police, to regard all Swedish passports as legal being held by any Jew in Hungary. Also, that these Jews couldn't be deported as they had always been before, and as over one million of them had already.

"It was these facts Eichmann and I discussed that day, as well as the dead Jewish bodies I'd seen floating face down in the Danube, that had put me in the state of mind to witness that being I still believe was a mirage. A state of mind, Dr. Grieg, which you yourself explained to Mr. Dillinger, might bring about a telepathic awareness from one's unconscious mind due to a feeling of paranoia. I had a lot to be paranoid about, especially after what Eichmann had said about having me killed whenever he liked."

"Yes, Mr. Wallenberg that might be true. But seeing an angel on that particular hill makes me almost certain you actually saw the real thing, and not something projected from out of your unconscious mind. But tell us how you arrived here; was it by a cloud? Or something no one has discussed as yet?"

"It was by a cloud, Dr. Grieg. And are you not at all surprised it could have been otherwise?"

"No, Mr. Wallenberg, I can't ... Where and when did you see this cloud before it suddenly enveloped you?"

"Do I really have to tell you the place I saw this cloud, Dr. Grieg? Is it not obvious?"

"You mean the *Hill of Roses* again?"

"Yes, exactly. Only this time I wasn't on that hill voluntarily. I was instead a prisoner of the invading

Russian army."

"But why did the Russians put you under arrest, Mr. Wallenberg?" we all heard Anastasia ask, as I saw Bormann then lift an eyebrow in response.

"It was because of a number of reasons, Your Majesty, first, because they believed me to be an undercover spy for the Anglo-American secret services. This they believed because I'd been educated in the United States as an architect and years later dispatched to Budapest to save Jews by President Roosevelt's War Refugee Board, after they'd learned I'd done business there for the company I was then working for in Sweden.

"Another reason the Russians had me arrested was because they didn't like what my rich uncles had been doing behind their backs. You see, my two uncles, not my father, who had died at twenty-three of cancer months before I'd been born, had inherited their vast fortune from their grandfather, who had made it founding the Stockholms Enskila Bank. This fortune my uncles had shared between them, had made them major contact points in the west for the German anti-Nazi underground, since one of my uncles was living and doing business in London, and my other uncle in Stockholm. And it was through their worldwide business dealings that they became a friend of the mayor of Leipzig, Germany, Karl Goerdeler. He'd been one of the first public figures to break with the Nazis in 1936 when he'd learned about their anti-Semitic policies. He'd later gone, in 1942, to see my uncles in Stockholm, asking them to act as a liaison between Winston Churchill and the German underground who wanted to get rid of Hitler. You see, Herr Goerdeler and those in the underground, wanted my uncles to give them advance assurance from the Allies that they would make peace and stop the war with Germany if they had Hitler killed. But as you know, Dr. Grieg, this particular plot to kill Hitler failed because of infiltrators inside the conspiracy. Herr Goerdeler was then arrested by the Gestapo and sentenced to death in 1944, but was kept alive as a prisoner by Heinrich Himmler, head of the SS, because he wanted to use Herr Goerdeler's contacts with my uncles and the Allies to make it easier for him and Germany after the war was lost. But my uncles told Himmler that the Allies wouldn't agree to any peace agreement at that stage of the war, especially when Germany was about to surrender. So Himmler then had Herr Goerdeler put to death, I suspect..."

"You're quite correct, Herr Wallenberg. And I enjoyed the sight," Bormann suddenly remarked with a smile, we all observed.

"Shut up you Nazi pig bastard!" Mr. Howard then yelled.

"Go on, Mr. Wallenberg...Just ignore him," I said.

"Yes, I will. Well you see, the Russians believed my uncles had been attempting to negotiate, without them, a separate peace with the Western Allies and Germany."

"That's all very interesting, Mr. Wallenberg. But how was it you were again on that *Hill of Roses* with the Russians that had arrested you?" I asked him again.

"Isn't it obvious, Dr. Grieg? The Russians were looking for Adolf Eichmann, and they knew I knew what he looked like. So they took me to his villa on that hill immediately after I was arrested. But little did any of us realize what would take place on the *Hill of Roses* only minutes after we arrived..."

"Please go on. Tell us what happened in those first few minutes?" I said, thinking again he had stopped speaking because of Bormann's presence. The presence of a man I knew Mr. Wallenberg detested being anywhere near, and I wondered why he had not as yet insulted Bormann to his face. However, one had only to look at Mr. Wallenberg to know that insulting anyone was a thing he would never do. It was too below him, too below the fine gentleman everyone knew him to be. One could not but notice a refinement of character that a man of his young age rarely possessed. And he seemed to know this about his aristocratic manner, the way he carried himself, and by the way he almost charmed you by his friendly voice. From what little I knew about him, in my time, I knew he had used all of his fine character traits to assist him in bringing people around to his way of thought. I also knew Mr. Wallenberg to be a master of the human psyche. Anyone who could have dealt with the likes of an Adolf Eichmann, and lived to tell of it, had to have been also a master of Freudian psychology, as well as a master in the art of negotiating a deal.

I then began to notice Mr. Wallenberg's face take on a look of sublime reflection, as if he were now going over in his mind the mysterious things he had seen happen on that hill. And I knew from this look, he was about to respond to my still unanswered question. As everyone around him watched his penetrating and inquisitive eyes then lock on to my face, we all heard him say, "Well, Dr. Grieg, what I and these Russians saw when we all arrived on that hill was like nothing any of us had ever seen. And I'm not referring to this angel, but rather something along the lines of what happened at the World War One 'Battle of Mons' that Mr. Cooper explained."

"You mean you and those Russians saw phantom armies on that hill?"

"Not exactly, Mr. Mallory. Nor did any of us see ghostly squadrons of cavalry riding along our line of march, nor knights, bowmen, or levitating Catholic saints either."

"Then just what, for Christ's sake, did you and those bloody Russians see on that bleeding damn hill, old chap!" Mr. Howard then shouted out in exasperation to everyone's surprise.

"Well, Mr. Howard, it was a bunch of half-naked wild and crazy cannibals!"

"You've got to be joking, Mr. Wallenberg? But then what happened? And don't try and tell us that those cannibals then ate up all those Russians either, because those cannibals of yours would have been shot down dead in their tracks by the many machine-guns the Russians must have been carrying."

"Quite so, Mr. Hoffa, but..."

Suddenly, everyone heard Mr. Rockefeller interrupt Mr. Wallenberg and say, "Hey! Wait a second! Doesn't anyone remember that cannibals were chasing me moments before I arrived in here?"

"Hey! That's right! Then how the hell did they get on that hill in Budapest in 1945? And just how many of them cannibals were there you two gents saw?"

"I think there was about a hundred of them on the hill, Mr. Howard," Mr. Wallenberg answered.

"That's right. About a hundred. But tell me, Mr. Wallenberg, did you happen to see a cloud on that hill before you and those Russians arrived?"

"Yes, Mr. Rockefeller, I did. But doesn't anyone want to know what happened to me and those Russians after those cannibals materialized?"

"Of course we do, old chap. But it's plainly evident to us that that cloud must have then left the hill with only you aboard. But were you fortunate enough to have seen the shocked and disbelieving looks on the faces of those Russians before you were lifted up and given that life-saving ride of yours?"

"Yes, I'm happy to say, I did Mr. Howard. They just about turned purple from seeing them, as I then watched them all hightail it back down the hill in a maddening rush. As to what happened to those spear-carrying crazy cannibals, well that's anybody's guess, but chances are they probably were all teleported back to where they came from the moment I arrived in here."

"Well let's hope so, Mr. Wallenberg, because I don't think we have enough bullets in these guns I collected to kill more than a third of them."

"Oh, Dr. Grieg, I think I should also add that there's another connection between Mr. Rockefeller and myself."

"Oh, you mean the fact that both of your families are super rich, Mr. Wallenberg?" I asked.

"Yes, exactly."

"By the way, Mr. Wallenberg," I said. "I think you'll be happy to know, that after your disappearance, Adolph Eichmann was later captured while living in South America by agents of the newly created nation of Israel, and then executed to death after his trial."

"That's a blessing to hear, Dr. Grieg, I'm happy to say," Mr. Wallenberg said, as we all then noticed Bormann give him a worried stare.

"Hey! I think I have something in common with you, Dr. Grieg!"

"What's that, Miss Earhart?"

"Well I studied biology at Columbia University as a pre-med student. You know I even have something in common with that man who disappeared on Mount Olympus."

"Oh you mean Mr. Macmillan, of the famous publishing house?"

"Yes, Dr. Grieg, because my husband is George Palmer Putnam, the owner of the Putnam publishing house."

"Wasn't he also once the head of the editorial board of Paramount Pictures, Miss Earhart? Or should I say, Mrs. Putnam?"

"Miss Earhart is fine, Mr. Howard. And yes, my husband did work for Paramount Pictures. He initiated a number of film projects, including the aviation movie epic called *Wings*. It's too bad you weren't there at the time. We could have become fast friends."

"Yes, I'm sure we would have hit it off well. Like

we're doing now, by Jove."

"Oh, I almost forgot, Judge Crater," we all saw her turn to him and say, "but I seem to have a connection with you as well."

"Oh, Miss Earhart, What might that be?"

"Well my father was a lawyer, and my mother's father was a judge like you."

"That's amazing," Judge Crater said.

"And then there's the connection I have with you, Monsieur Saint-Exupéry."

"And what is that, Mademoiselle Earhart,. other than the fact we are famous aviators?"

"Well, Monsieur Saint-Exupéry, I also studied French at Columbia University, and became as a result, fascinated by French poetry. And of course there's the connection I also have as a pilot with you Mr. Kennedy."

"Yes, but I think Mr. Amundsen is also a pilot as well," Joe Jr. said.

"Yes, Mr. Kennedy, you're correct. I was the first person in Norway to obtain a civilian pilot's license," Mr. Amundsen replied.

"Miss Earhart, you seem to have a connection with me," I said. "Poetry is also one of my loves, especially the poetry of Walt Whitman. And I also seem to remember that your plane, the plane you were flying before your disappearance into this place, had the name of a Greek goddess, or maybe some princess I believe?"

"The plane I was flying, Dr. Grieg, was a *Lockheed Electra*. And *Electra* was the ocean nymph and mother of the goddess *Iris*, who was the goddess of the rainbow and the messenger of the gods. Why do you ask?"

"Well, I think this angel some of us have seen so far, may very well be a goddess of some kind known throughout history by many different names. But I'm not ready at the moment to explain my theory about this now. At least not until I hear everybody speak."

"Dr. Grieg, I think you would be very much interested to know what the Tibetans call Mount Everest, that 29,000-foot mountain which my partner and I almost conquered in 1924, the year of my third attempt before arriving here."

"Yes, Mr. Mallory, I would. If you think it's that important. But weren't you the man who said, when asked why he wanted to climb Mount Everest, whimsically replied 'Because it's there', if I'm not mistaken?"

"Yes, Dr. Grieg, I'm the guy. But let me tell you what the Tibetans call that awesome mountain at the top of the world in their language: they refer to it as *Chomolungma*. The English translation of that word is *Goddess Mother*. Now isn't that amazing after what you just said, Doc? I mean, talk about connections?"

"Yes, it certainly is an amazing one at that. But just how did you get here from that location? Was it a cloud that was responsible, like some of us already mentioned?"

"Well, not exactly. It was actually a fine mist, my partner, Mr. Irvine, and I saw at about 28,000 feet, Dr. Grieg. It was immediately after both of us saw this mist that it became so thick we couldn't see even a bloody foot ahead. And then, not more than a minute later, I found myself in this foggy flat plain very much dumbfounded and alone. It was then that I decided to ditch my oxygen tank I had strapped to my back."

"Did you happen to see an angel, or being, like some of us have seen?"

"I was just about to tell you that, but maybe I should first explain to you and the other people here, why the Tibetans call Mount Everest the *Goddess Mother*."

"Yes. Please do," I said, nervously waiting for his reply.

"Well the Tibetans, as well as the Hindus, believe that at the top of the world there exists an area where the Mother Goddess of the Earth makes her home. Now I don't mean at the top of Mount Everest, Dr. Grieg, but somewhere directly above it in the sky."

"You mean like some sort of 'astral plain', or something to that effect?"

"Yes, exactly, Mr. Howard. A place that touches closest to the earth and the heavens; a place where, if such a being as this 'Mother Goddess' exists, can communicate with the least effort with man and the Lord God Almighty...Oh! And I might add, the Tibetans also believe that at the summit of 'Su-Meru', the mythic Mount Sinai of Tibet, dwell the rest of the gods. And that these gods live to be one thousand celestial years, equaling thirty-six million of the years of man...But listen to this, Dr. Grieg, they also believe at the summit of Su-Meru, one day is equal to a hundred years on this earth. If that's not the best possible theory of where we are at the moment, then I don't know what is."

"You may be right, Mr. Mallory," I said. "Because if we're all in such a place as you've just explained, then that would account for why we're not all hungry. And why time seems to be out of whack in here. The reason I say this is because at altitudes above ten thousand feet, a person begins to lose their appetite, due mainly because of the negative ions and gases they are breathing. As a matter of fact, there are some yogi masters of Tibet I've read about who because they study 'Hatha Yoga' have learned to live off only the air. They can do this only at high altitudes, but only after first converting, over the years, from a solid food diet, to a liquid one and then to no food at all. They can only exist without food at the highest mountain levels.

"And as far as time is concerned, well, if we're someplace above Mount Everest, or maybe this 'Mount Su-Meru', then perhaps, because of these two mountains being somewhat shaped like a four-sided pyramid, and their tremendous height, time as we know it is being affected by maybe a magnetic anomaly of some kind, much like in the area of the 'Bermuda Triangle' possibly. Then again, we may be all in an altered state of consciousness where time seems to have come to a complete halt. For example, has anyone ever glanced at a wall clock or their wristwatch and observed the second hand not moving for what seemed to be a few seconds? Well whether anyone realized it, you were at that moment in another kind of time, a time where your consciousness was operating at another brainwave pattern. So maybe that is what's happening to us here: we're all in a state of altered consciousness, caused by that cloud up there maybe. It might very well be causing some kind of time dilation for all we know.

"Still, like I have repeated, not until I have heard everyone speak of their experiences before arriving here will I be able to come to any definite conclusion on what's happened to us...Mr. Mallory, you still haven't told us if you saw an angel at any time in your life?"

"Well if you want to call it that, Dr. Grieg. I did see something of a celestial being as I was climbing Mount Everest. But one sees many things before him at altitudes above ten thousand feet. One can become so lightheaded on reaching that height one can even have what is called: 'an out of the body experience'."

"Pardon me, Mr. Mallory, but did you have a brother by the name of Trafford Leigh Mallory?" we all heard Mr. Miller then ask.

"Why, yes I do, Mr. Miller. But what do you mean by 'did' I have a brother? Am I to understand my brother, Trafford, died after my disappearance here?"

"Well I hate to be the one to inform you, but it was in November 1944 that your brother died. But he was at that time, Air Chief Marshall Sir Trafford Leigh Mallory, senior commander of the Royal Air Force in World War Two, and the highest ranking British officer to die in the war."

"Mr. Miller, just how was my brother killed?" everyone heard Mr. Mallory ask with a saddened voice, his eyes on the verge of tears.

"He was unfortunately killed when the plane he was traveling in crashed into the French Alps during bad weather. Everyone on board the plane was killed, including his wife. He had been en route to Burma when this occurred. The reason I know about this is because I was in England a month after this happened, the very month I myself disappeared."

'Oh, I see...I am grateful you told me this about my brother, Mr. Miller, even though it's so painful for me to hear. I always knew Trafford would advance to the top of his profession. He'd joined the Royal Air Force very young in his life...You know, we came from a family of Rectors going back to 1621. My brother and I were the first not to make the church our life. I became a history teacher and a diehard agnostic instead, even though I attended Mary Magdalene College and was taught by a history professor who was also the son of the Archbishop of Canterbury....By the way, do you happen to know if my wife, Ruth, and my two children are alive and well in your time of December 1944?"

"Yes. I believe they are, Mr. Mallory," Mr. Miller said, not knowing that Mr. Mallory's wife had died of cancer in 1942 from what I had learned on a TV documentary of his life.

"Well that's at least comforting to hear," we all heard Mr. Mallory then answer.

"Mr. Mallory, I'm sorry to hear that about your brother, but I must ask you if you saw anything looking like what has already been described by five of the men here?"

"Well, yes. But like I said, who knows if I was only hallucinating, Dr. Grieg."

"Well, no matter..." I said, as I turned to Miss Earhart and asked, "Miss Earhart, I believe you were telling us about some of the connections you seem to have with us. Would you please continue now by telling us how you happened to get here?"

"Well you see, gentlemen and Your Royal Majesties, as part of my flight to be the first woman to fly around the world at the equator in July of 1937, I was at the same time on a secret mission to determine if the Japanese were in violation of their League of Nations mandate. If they were in fact re-enforcing and supplying arms to their bases on the Micronesian Islands, as well as the Mariana, Caroline, and Marshall Island groups, which had been previously governed by Germany. It was while I'd been flying over Howland Island, just west of the International Dateline, and within a few miles of the Equator - Howland Island being the island I was going to land on after flying almost three thousand miles without a stop and departing Lae, New Guinea, that my plane began to enter a mysterious white haze. This made it impossible for me to differentiate between the water and the horizon.

"You see, human sensations fail one when one is blind. When one is deprived of a horizon a flyer may lose the feel of his position in space. At times like these, most pilots can't tell if they're flying with one wing high, or perhaps turning. Pilots just can't be sure, one way or the other.

"So as I flew into this white haze, I tried to keep my plane in flying trim, with my one wish at that moment growing stronger and stronger: to see the friendly looking earth again...Ordinarily in cases like these, I would have pulled back on the stick and kicked the rudder, thereby sending the plane into a spin, which was an unwise and dangerous thing to do, for it could have stalled out all the engines.

"But before I could even think of this maneuver, the entire plane began to shake from its tip to its tail. And it was only a few seconds later that my plane's two engines did stall, leaving my copilot, Freddy Noonan, and I, with the only thing left to do before crashing our plane five thousand feet into the water below."

"And what was that, Miss Earhart?" I asked.

"To pray, Dr. Grieg that the plane's two engines restarted themselves again because we carried no parachutes aboard. We reasoned, if we crashed out over the Pacific Ocean, parachutes weren't going to do us a bit of good anyway."

"Then how did you survive this crash, Miss Earhart?" everyone heard Anastasia ask.

"I was just about to tell all of you how... Well, believe it or not, both Freddy and I suddenly found ourselves being sucked right through the blown out windows of the plane's cockpit as we both were trying to desperately restart our two failed engines. How our cockpit windows got blown out, from maybe the outside atmospheric pressure, is a mystery to me, as is what then happened to Freddy and I immediately after this disastrous event. As we suddenly found ourselves outside the plane, watching it as it spiraled out of our sight through this thick and blinding haze, we were dumbfounded to then find ourselves being carried by the wind into a glowing cloud directly above.

"As I called out to Freddy as this was occurring, I realized we'd lost sight of each other the moment we made it into this cloud. And no matter how much I yelled out to him during that frightening time, I couldn't get him to answer.

"Then instead of passing through this cloud, as I knew I was going to do within seconds of entering the thing, I found myself mysteriously being lifted upright, then gently lowered on to what I knew to be a solid flat-surfaced floor of some kind. I thought at that instant I'd gone insane, because I knew without doubt I was still inside this particular cloud. However, after landing feet first on this floor, I had no other choice but to believe I was standing on actual solid ground. And it was then I knew I was still alive, as impossible as that seemed.

"Then in what appeared to be only a few seconds, I suddenly heard in the foggy distance what I recognized to be a trombone being played and decided to take a chance and venture off towards where I believed this blaring sound was originating."

"Let me ask you now, Miss Earhart, what I have asked six of us already..."

"Yes, I know, Dr. Grieg. Did I see an angel before or after I entered this place? Well yes! I did! But during the two times I saw this being, I couldn't really tell if I'd only been hallucinating her presence or if she was in fact real."

"You mean you saw this being more than once in your life?" I asked her, a little bit spooked.

"Yes, immediately after the two crash landings I had in my earlier years: the first crash, occurring on a farmer's field during a thunderstorm. I'd been lucky in that one. I'd been left hanging upside down from my safety belt after my plane nosed over onto its back. And in my second crash, I'd been forced down by a downdraft, flipping my plane over in a field of overgrown weeds, where there was a strong smell of roses in the air. Only in this crash, I'd been flung out of the cockpit because of my safety belt suddenly breaking the instant the plane flipped over.

"Anyway, because of the fact I'd been badly shaken up by these two crashes of mine, I suspect my brain was in such a state that I couldn't help but see stars before my eyes.

"However, it was only seconds after seeing those illusionary stars that I also saw, appearing before me, this strangely vaporous looking woman. In the first crash, I saw her upside down because of being upside down in my cockpit, and in the second crash, fortunately seeing her upright. But I don't really remember seeing her for more than a few brief seconds. This was probably because I began losing consciousness after my two crashes almost immediately. So don't ask me to describe what this being was wearing, Dr. Grieg, because I have absolutely no recollection of her attire at all. But of her face, well that's something I will never forget seeing. She had a face only an angel could possess. If in fact she was an angel I had the privilege of seeing."

"What did her face look like, Miss Earhart? Can you please try and describe it to us," I asked with great eagerness.

"Well as far as I can put this into words, her face was of a woman in love, if you know what I mean. She had an expression of great compassion, the kind of look you might see on the faces of those saints depicted in paintings of the Middle Ages; and a face that had the look of a peace that passeth all understanding, as the Bible so rightly says. It was a face I knew to be of a being not of this world: a being who instead could only have come from a place where the Lord's holy angels make their home. And if that place is what we all understand to be paradise, then that's where she must have came because I can't possibly think of any other place in this world, excluding here where she could have actually originated. But does any of this mean anything to you, Dr. Grieg, anything at all?"

"More than you can realize, Miss Earhart. And may I say you are quite a courageous young woman. I mean to be the holder of so many world records. You make Lindbergh look like an amateur because not only were you the first woman to have flown solo across the Atlantic, as most of us here are already aware, but you, as I remember it, set at least a number of other records on top of this remarkable feat."

"Oui oui, Dr. Grieg, Madame Earhart is an 'aviatress extraordinaire' for she was also the first woman to fly across the Atlantic as a passenger before soloing across it. And the first person to fly solo from Honolulu to California, from Mexico City to New Jersey, and to have set the record for autogiros at 18,451 feet. These are truly remarkable records, Madame Earhart. Truly 'formidable', as we Frenchmen say."

"Why, thank you, gentlemen. You're so kind. But Monsieur Saint-Exupéry, you too are quite remarkable. Because besides being a great aviator, you are on top of this, a very great writer. Your novels about flying, which I've read, are loved the world over. And did you know, Dr. Grieg, Monsieur Saint-Exupéry is called the 'Joseph Conrad of the Air' because of his two most memorable books?"

"Yes, I know, Miss Earhart. But that is equally as true with the other books Mr. Saint-Exupéry has written since your disappearance. And speaking of one, wasn't that book *The Little Prince* you wrote, Mr. Saint-Exupéry, about a pilot who crash landed in the Sahara desert, and meeting a little prince soon after, the only inhabitant of a far distant planet, very much like what has happened to us here?"

"Oui, Dr. Grieg, it is. But you're forgetting to make mention of the fact that this book was also about this little prince's love of a rose, of his devotion to this flower that radiated through his entire body and mind like the rays of the sun ... It is this love he has for this flower, the only one on his tiny world that connects my book about him with the smell of roses in this world, if not our very predicament. I cannot understand why this should be so, Dr. Grieg, but it is truly an extraordinary set of circumstances, is it not, mon ami?"

"That's for sure. And we even have a little princess among us as well, if that was what you were referring to, Mr. Saint-Exupéry?" I said, truly amazed.

"Oui, Dr. Grieg, among other things, but maybe it is my turn now to tell you of my seeing the angel, is it not?"

"Yes, Mr. Saint-Exupéry, why don't you tell us about that next," I said to him with a nod.

"Well, as you said, Dr. Grieg, I based my story of *The Little Prince* on my own personal experiences. You see, I once had the misfortune of crash landing along with my mechanic, Andre Prevot, in of all places, the god-awful Sahara desert, as had the pilot in this book of mine.

"As to why Prevot and I had been flying over this vast hot hellhole of a place, we were attempting to break a speed record of eighty-seven hours and win the grand prize of fifteen thousand francs offered for flying from Paris to Saigon, on that fateful day of December 29, 1935.

"Anyway, after landing our plane at Marseilles, and fixing a leaky fuel tank, we then set off out over the Mediterranean and across Sardinia for Benghazi to refuel, and then on out to our next destination of Cairo.

"We had planned on flying along the Libyan coast, but because of the Abyssinian war, we had decided to fly midway between Alexandria and Cairo for the six hundred miles of desert ahead. But on climbing to the height of six thousand feet, where we thought the winds were likely to be most favorable, we instead entered the likes of a towering mass of cumulus clouds forming eerie halos around them in the mist. And when we then descended to about twelve hundred feet, some four hours later, changing our course toward the north, hoping to glimpse the sea and the bright lights of Cairo to our left, we were shocked to observe nothing but complete and utter darkness.

"However, after I had banked the plane to the right on a north-northeast heading, in order to avoid another massive cloud, I suddenly heard Prevot cry out, at about nine hundred feet: 'A lighthouse Antoine! A lighthouse!' And it was in that moment that we heard a frightening loud and eerie shudder throughout the length of the plane, followed instantly by a dizzying rotation as one of the wings struck the ground of a high, hilly plateau. It had been made up of many round black pebbles and had the effect of cushioning our impact, acting like ball bearings beneath the plane's fuselage, which I later realized had resulted in saving our very lives.

"Still, it was only moments after we survived this crash, walking away from what was left of our plane that we knew we were not going to be able to radio for help, for we had earlier decided to dispense with a wireless in favor of adding extra fuel tanks in the plane instead.

"I later wrote all about this crash experience and near fatal ordeal of three days in my book, *Wind Sand and Stars*. The book won the Grand Prix of the 'Academie Francaise' and the American Book Award, and was an international bestseller in 1939."

"Mr. Saint-Exupéry, please tell those of us who have never had the opportunity of reading that book of yours, how you and your friend got out of that deadly hot desert alive," everyone heard Miss Earhart ask.

"Yes, that's a good idea," I said. "Because your answer to Miss Earhart's question, Mr. Saint-Exupéry, might just help us find a way out of this similar no man's land."

"Well, Dr. Grieg, what you want me to relate now has to do with that angel I was in the process of telling everyone about. And you know, having never written about seeing this angel in the desert in any of my books, or spoken to anyone about this other than my friend Prevot because of the ridicule I knew I would have received, I find it quite comforting I have an audience of people who have seen an angel and cannot but believe in me now.

"Anyway, my friends, of the five near fatal crashes I have had, this one in the Sahara desert was the most life threatening to me, if not the most memorable in my experience. As to how Prevot and I escaped out of that desert alive with the help of this angel, well, it is a very long story. So to begin it I will tell you, that this, my third crash, after the first few hours of it, I began to head off with Prevot toward a ridge some twenty miles away. However, on reaching it and climbing to its top, we were unable to observe anything except mirages which disappeared on our approach. So we immediately decided to return back to our plane, figuring we would have a better chance for an aircraft to find us. And it was there where we quickly built a great pyre made up of bits and pieces of the splintered wing, gathering together right afterwards, what was left of our gasoline to throw on this pile of wreckage when the day became night.

"It was at that time, however, that I became strangely haunted by a vision of my lovely wife's eyes, while poor Prevot began expressing what little hope he had for our rescue by openly weeping. I guess it was because we had just run out of what little water we had brought along in the plane. Then after asking him what he had been so openly crying about, I remember him replying, 'Do you really believe I am crying about myself, Antoine?'

"Well anyway, at daybreak, Prevot and I managed to get a little water off of the plane's wings because of the morning dew that had formed on them, the water tasting like hell, as I remember it, because of the small amounts of engine oil and airplane paint that had worked their way into it. But at least we had obtained the needed moisture to wet our dry and cracking lips.

"Then, on this, our second day, we began to build yet another bonfire in case a rescue plane should happen by. But on that day nothing was to be heard overhead. So when daybreak arrived, I then decided to go off alone into the desert to find perhaps some food, water, or help. There was no sense for Prevot to then accompany me, for I knew I might die out in the hot blazing sun a whole lot sooner then him because of the lack of any shade. And I realized Prevot might be seen more easily because of the wreckage of the plane by remaining behind.

"However, it was two hours later into my desert walk that I remember having my first hallucination of what I thought to be the faint outline of a man. Yet after I had yelled out and made motions in the air with my arms trying to get his attention, I suddenly realized this man had only been a big black rock, and less than thirty minutes later the same thing happened again, when I mistook the trunk of a tree for that of a sleeping Arab Bedouin. Then, as I quickly realized there could not possibly have been a tree recently growing there, I remember bending down to examine it, only to be amazed when I found it to be only a solid piece of petrified wood. As I looked around at that moment, I saw that this was only one of many other pieces of petrified wood just a few yards away. And I knew long ago into the past, an antediluvian forest had once existed at this spot.

"It was then that I saw ahead of me what I believed to be a passing caravan making its way across the horizon. So I began to shout out to it, but to no avail, I did not get anyone to answer. I again realized I had only been imaging what I more than desperately wanted to see.

"Yet as if I were still under some kind of hypnotic trance from the effects of my dehydration, I saw about ten miles away, and on the top of a hill, what I could only presume to be the configuration of some giant cross. And it was then that I remembered seeing on a map the night before, a circle and cross on the map representing some religious order.

"I also remember seeing on the map, a single black dot beside this circle and cross, representing none other than a permanent well. So you can imagine how overwhelmed I was at that moment when I realized what was now up ahead. Not to mention the fact that there had also been two white circles next to the black dot on that map representing two temporary wells. So I figured I couldn't possibly fail to find water in one of the wells when I finally arrived.

"Yet I was to be disappointed again. The cross on the hill turned out to have been an illusion. And the circle and cross I had earlier seen on the map, I had sadly confused to be in the area in which I was now standing, since I could not have possibly known my correct position then in that empty white wilderness of sand.

"However, I was to have yet another hallucination during this unhappy event. For I then saw an entire city over the hilltop at the horizon, a city of such beauty, that I had at once realized it to be nothing but a mirage. And I knew once again, I had been probably suffering from the effects of the insufferable heat, knowing not until the twilight mercifully arrived, would I get any cooling relief.

"When the twilight finally appeared, I at once realized how far I had ventured off from the crashed plane and poor lonely Prevot, thinking then, if it was at all wise to go back to him. I knew Prevot might already have been rescued by some passing caravan, and what lay ahead of me could very well be the help I had been seeking.

"Still, as I pondered this over in my mind, Dr. Grieg, the more I realized I had but one course to follow. And that was to turn back the way I had come, for I couldn't be sure Prevot had actually been saved.

"So after walking for more than three additional hours back to him and the crashed plane, I was stunned, not to mention overjoyed, when I observed standing in the light from the bonfire Prevot had already lit, two Arabs talking to him. And as I immediately began yelling out to Prevot as the two Arabs turned, Prevot following suit, I saw him begin to make his way to meet me only a hundred yards away. But after I had asked him about when the two Arabs had mysteriously appeared he could only look at me perplexed and say, 'What Arabs, Antoine? There are no Arabs here. You must have been seeing another mirage.'

"Nevertheless, not letting this letdown get the better of me, I immediately told Prevot to rip up one of the parachutes and lay portions of it on the desert floor, weighing them down with some available stones. I reasoned, the portions of parachute would by morning, be moist with dew, as had the plane's wings, giving us perhaps enough water, after wringing the chutes out, to satisfy our longing thirsts.

"It was at this time, however, when Prevot had been getting an empty jerrican to contain the expected water we might obtain by morning that he suddenly shouted to me from the wrecked plane, he had found an orange. We shared it immediately, and afterwards told each other of how wondrous it had tasted, and the joy and soothing relief it had given to our parched and swollen mouths.

"When by morning we had been fortunate enough to have collected quite an enormous quantity of water from our effort: almost two whole quarts we had then carefully poured, we were disappointed to discover we had filled the water into an empty and unwashed jerrican, thus making the water green and yellow and smelling like hell. And after we had almost thrown up from drinking a small amount of this unpleasant brew, a brew that had tasted of rusted metal and gasoline, poor Prevot began to wildly spin around as we then both vomited up what we had already consumed. I, as I remember it, going through a series of convulsions too, this probably being caused by the magnesium lining of the jerrican I could only suspect at that most sickening time.

"However, by morning we had decided to break the pilot's unwritten law of sticking to a crashed ship by heading off into the desert for help, we wanted to get away from this cursed plateau, not believing anyone was going to look for us there. We were in such a worried state over what our loved ones must have been going through, that we could not remain idle waiting for someone to come - Prevot even saying he would have laid down and died had he not anyone back home to go to that he loved, and I saying the same.

"Still, of that day, having ventured off from our crashed plane, I can remember nothing but that I had walked most of the day with my eyes to the ground because of the mirages I feared I would see, knowing I would not be able to endure yet another defeat. But of the night, I do remember Prevot and I again placing the pieces of parachute on the desert floor in the hopes of catching the morning dew. However, it had been a cloudless morning and the dew we had prayed for had not chosen to materialize as we had so greatly wished.

"When later that day, after I had watched Prevot suddenly stop dead in his tracks because of the unbearable heat and say to me he could see a lake some distance away, I remember calling him crazy. He then began telling me if I had ever heard of a mirage appearing after the sun had completely set. As I reacted to that bit of his logic by remaining silent, unable to answer, I knew I had finally given up believing in my own eyes. As Prevot began taking my silence as a sign for him to go off by himself, some twenty miles toward this lake he alone could make out and see, I began to shout to him in an emotional outburst, telling him to go ahead and take his foolish walk to his lake, warning him it would only be filled with salt and too far away.

"And as Prevot began walking off knowing I could do nothing to dissuade him, knowing I would not follow in his foolhardy steps, I took the opportunity, figuring he would return shortly, to write a last loving letter to my wife. I feared my letter would have a posthumous delivery all too very soon.

"Later that night, I was suddenly dumbfounded

to see, about five hundred yards away, what looked to be the light from a brightly glowing lamp. And it was then that I started wondering to myself if perhaps Prevot had lost his way and was now returning to find me in the dark. I knew he had taken the only lamp on his journey, preventing me at this time from signaling him. So I immediately began to stand from my laying position and shout out to my friend, when to my amazement I suddenly saw appearing next to his lamp, yet two more swinging near its sides. Could it be possible, I had reasoned, that a search party was coming to rescue me now, together with my friend Prevot? So I began shouting in their direction at the top of my lungs to get their attention, realizing as I did, they were not hearing me no matter how loud I was frantically shouting. Then running to the lights I could still see, yelling at the men I believed were holding them, yelling to, 'Wait! Wait!' I saw to my horror the light of their lamps turning away from me in the dark.

"Then too exhausted to continue, knowing they had heard me, and knowing they could not have failed, I stopped in frustration to regain my breath, when to my surprise, I heard someone calling out to me, as if whoever it had been was now returning all of my unanswered cries.

"As I began running in the direction I believed this man's comforting voice was coming, I suddenly found I had run straight into the arms of Prevot, where exhausted and knelling over, I said to him I had seen two other lamps. Then as if totally bewildered by what I had said, I heard him reply he did not know what I was so wildly saying, and I realized I had been seeing yet another mirage.

"Then believing Prevot had come back to me after discovering what I had thought to have been just an imaginary body of water, I remember asking him with an alarming voice if he had found the lake he had so fervently seen.

"However, after hearing him sadly explain the lake had kept moving away from him as fast as he could follow, then deciding to come back to me even though he still believed there was a lake still further ahead, I remember berating him for risking his life chasing after what he could not have known had been there.

"And finally, as he explained he had only gone off looking for water because of how terribly pale I had seemed, you can imagine how I was feeling as I remembered what I had earlier said to him about being out of his mind. Knowing Prevot had been only half-crazed at the time of his trip, I realized he had only been concerned with my health. So after replying to him moments later, looking back into his warm and friendly eyes, of how positive I had been about seeing those other two lights, I saw him look back at me with a seriousness I had never seen before. On hearing him reply in a voice quite quivering with fear, 'Antoine, we must both be out of our minds,' I could only agree with him. Because later, after we had gone to sleep, I had found myself awake and on my feet shivering, unconsciously venturing off and away from him and wondering where the hell I was going.

"All the same, I thank God I had heard him shouting to me in the dark at that time, awakening me from out of my delirious stupor. Yet as I made my way back in his direction, freezing and in a fit of hiccups, understanding myself to be suffering from only the effects of dehydration, I began heading away from him again, and heading instead for our first-aid kit containing what I knew would be bottles of alcohol, pure ether, and iodine. As I began reaching for the bottle of ether, finding myself in the process of stupidly swallowing some, and reaching for the bottle of alcohol and swallowing it as well, I remember digging a hole in the sand and laying myself down into it to keep myself warm, covering my entire body until only my head could be seen. As I curiously watched Prevot begin a small fire out of some nearby twigs, and watch him later stamp around in a circle when his fire had burned itself out, only having to cover himself with sand as I myself had already done, I nearly wished I would die in my sleep in this hole I had dug, rather than awake in the morning to later die in the hot blazing sun.

"Because by morning, Dr. Grieg, I knew I would awaken with a painful rasping in my throat, my tongue feeling like it was plaster of Paris, a horrendous taste in my mouth, and my stomach burning inside of me, due no less to the acid buildup it would be going through because of the lack of any food. I had eaten during the previous three days, only a few grapes, half a shared orange, and some stale hardened cake.

"When in the morning our ripped up pieces of parachute were bone dry, as they had been when we had awakened the morning before, we could not travel more than five hundred yards that day without our legs giving way.

"The day before, we had walked on with at least some kind of hope, but during this day we went on without it, for the word hope had now become quite meaningless to us without any water.

"Then just as we were about to give up and die where we were so sadly standing, in this, our greatest moment of despair, we both experienced seeing something that neither of us alone would have ever believed. When Prevot began describing to me exactly what I myself was then seeing, I knew without the slightest of doubt that we were both seeing something that was very real. When two people see the same thing in the desert, after not having agreed on seeing what the other had claimed to have seen, one is at once dumbfounded into realizing one is not seeing a mirage.

"Appearing before our astonished eyes now was what could have only been an angel, hovering in the air about ten feet above the sand and twenty feet from our heads, dressed in white - a luminescent white, with a brilliant aura surrounding her. And as we watched this angel then take on an attitude of prayer holding a string of white beads ending in a sparkling white crucifix, we instantly knew she was anything but a mirage."

"That's quite a compelling story, Mr. Saint-Exupéry. Then what happened?" I said, almost too mesmerized by his incredible narrative to interrupt, the faces of all the others around him seemingly as spellbound with wonder as myself.

"Well, Dr. Grieg, ladies and gentlemen, in what seemed to have been no more than a few seconds, after Prevot and I saw this miracle of an angel, we saw her raise up one of her hands - the one holding the white beads and crucifix, and point toward a sand dune not more than a hundred feet away. As we traveled to that dune being accompanied by her glorious presence, we were amazed to then find the fresh footprints of two men who had stood together, and the hoof prints of two camels that had just kneeled.

"As Prevot and I looked up at her most holy and beautiful face, a face like no other we had ever seen, we saw her raise her pointing hand away from the dune and out toward the eastern horizon. We immediately understood this action of hers to mean only one thing, and that was to travel the way she was pointing. Because the foot and hoof prints we might have followed on that dune were nowhere else to be seen, as strange as that perplexed and bewildered us.

"Stranger still though, was the discovery of a single white rose in the very center of the footprints we had found. How this rose could have existed in this desert was a mystery neither of us could ever explain, and we surmised the two men who had stopped on this dune had only done so because of this rose. But as to where the two men and camels had gone off and disappeared to without leaving any tracks, well I can only now theorize there had been a recent sandstorm near that dune, or they had been taken up in a cloud, as we have ourselves.

"As to what happened to Prevot and I after venturing off in the direction the angel had been pointing, well it was no more than a hundred feet into our walk that I heard what I recognized to be the crowing of a rooster. I thought for a moment I was only hearing an auditory hallucination at that time. However, when Prevot then grabbed me by the arm and asked me if I had just heard a rooster, I knew I had not been imagining that series of crows.

"Nevertheless, it was immediately after this I did have a hallucination. When I asked Prevot if he had just seen three dogs chasing one another in the distance, he began shaking his head at me and saying I was seeing another mirage.

"It was less than a minute later, however, we suddenly began shouting and waving our arms in the air, as we simultaneously saw up ahead what we knew to be the likes of an Arab Bedouin riding a camel. Still, try as we might, our shouts for his help seemed to be carrying only a mere thirty yards - much too short in distance to where he was slowly departing. Just as Prevot and I were about to give up in despair as we saw him fading from our sight, knowing he had been our last and only hope, we suddenly saw coming from behind the same dune, yet another Arab Bedouin riding a camel. Would this second Bedouin hear our desperate cries? We were wondering. And when it looked like our shouts to him were going to be unsuccessful, as they had for his unhearing friend, we saw him slowly begin to make a right angle turn in our direction, knowing instantly, when he came within fifty feet of us, our thirsts, the mirages, and our struggle to survive were to finally end.

"And end they finally did. For on approaching us without saying a word, he began placing his hands on our shoulders and laying us down on the sand. Then using the feather of a bird to open our dry and swollen lips, he gently began pushing a lentil puree into our dry thirsty mouths, so as to moisten our tongues and palates back to what was normal.

"And it was later we learned he had done this so we might absorb a mouthful of his precious life giving water, which, if we had drunk without first being given his lentil puree, would have only resulted in splitting our mucus membranes and causing our deaths. And die we almost did. For immediately on giving us a taste of his water, he had to instantly pull Prevot and me back to prevent us from suffocating and drowning in a wild and thirsty rage. Only after he had lifted us onto one of his camels and taken us to a nearby camp, did we ask him to go in search of more help, which by evening consisted of a car carrying four additional Bedouins. After setting out together for a thirty-minute drive to the house of a Swiss engineer named Raccaud, who operated a desert soda factory near some sodium saline deposits, Prevot and I were by midnight in Cairo, our original destination, thankfully alive and well and no worse for our wear.

"Nevertheless, here I am once more, Dr. Grieg, lost in another wilderness, a wilderness where I pray my beautiful angel will come to rescue me again."

"Yes, Mr. Saint-Exupéry, let's pray she does rescue us," I said. "All the same, let me ask you if you ever found out the reason for having that vision of your wife you said you had on the first day of your crash? I mean had she been sending you her thoughts telepathically, by chance?"

"Well it's funny you brought that up, Dr. Grieg, because all my life I have always been interested in magic, divination, thought transference, and hypnosis. I had even practiced hypnosis on my three sisters as a boy. These interests I had for the occult had been what attracted me to my wife; she had shared the very same beliefs in these things before we married. And on the day of my desert crash, I later learned, after being reunited with her, she had been at a séance trying to send me a message telepathically and because I had been in a receptive mood at the time, I had picked up the image of her face quite strongly."

"But did you pick up any of her words…the words she had been sending you, Mr. Saint-Exupéry?" everyone heard Miss Earhart ask.

"Yes, I did. But I have forgotten her exact words…something concerning my location at the time, I believe."

"You see, the reason I just asked you that question, Mr. Saint-Exupéry, is because my husband and I have always had a near obsessive interest in extrasensory perception. And it's a pity you couldn't have sent your wife a telepathic message in return. You could have told her your approximate location after your crash, because that's exactly what I would have done mentally with my husband."

"Yes, Madame Earhart, maybe I should have."

"Well it would seem, Mr. Saint-Exupéry," I said, "that your interest in telepathy with that of Miss Earhart's very much connect you with each other. And that's a very interesting connection you both seem to share. I wonder if we can use this ability in getting us home."

"But, Dr. Grieg," Mr. Saint-Exupéry was saying, "we do not even know where we are. And even if we did know, how would anyone come to our rescue but by a cloud like I myself jumped into while my reconnaissance plane was being downed?"

"Well, do you remember what I told everyone about my friends coming to look for me, after they realized they might use my backup vehicles to duplicate the process causing me to materialize here?"

"Oh oui, Dr. Grieg, I do remember you mentioning that. But then why would you need Madame

Earhart's and my telepathic abilities?"

"Well Mr. Saint-Exupéry, I was thinking more along the lines of getting help from that angel. The angel you said helped you and your friend in the Sahara desert, which I suspect is the same angel that has helped us all."

"Hey! That's a great idea, Doc. But isn't she responsible for getting us here in the first place?"

"We can't be certain of that, Mr. Hoffa. But do we really have any other choice in the matter?"

"No, I guess we don't, Doc," Mr. Hoffa was saying. "But before we try sending for her help via ESP, I think it would be a better idea to continue asking these people yet to speak about how they arrived. Not to mention, then describing to us their experience with the angel because maybe she doesn't want us to get out of here, at least not until we've all had our say."

"Yes, I'm beginning to see your point, Mr. Hoffa," I said. "Maybe we're here to work something out, something that might have to do with our immortal souls, perhaps?"

"You mean we might be able to get back to our times after going through this 'group therapy rap session' of sorts, Dr. Grieg?"

"Yes, perhaps, Mr. Kennedy, just perhaps. So who's the next to speak here?" I said, as I turned to Mr. Mallory and asked, "Mr. Mallory, can you please try and describe to me more fully your angel sighting on Mount Everest, if you will?"

"Well, Dr. Grieg, it was when Mr. Irving, my companion, and I were about 28,000 feet up the mountain that I suddenly saw her standing on a ledge above me. What happened after that you already know. And it's my guess I saw her because of what I told you the Tibetans call Mount Everest."

"Oh yes. You mentioned they refer to this mountain as 'The Mother Goddess.' Perhaps in your case, you were only hallucinating her presence, like you said:

the expectation of seeing her because of what the Tibetans called the mountain having an affect on your mind. Like a leading question from a hypnotist. If that's what you were trying to say, Mr. Mallory?"

"Yes, it was, Dr. Grieg. On the other hand, after hearing what happened to Mr. Saint-Exupéry, I'm not too sure I was hallucinating her now. Or for that matter, the smell of roses I just remembered having smelled only moments before noticing her...Oh. By the way, Dr. Grieg, do you happen to know if anyone after my unsuccessful climb ever reached the top of Mount Everest?"

"Yes, somebody did. His name was Sir Edmund Hillary, a New Zealand mountain climber and Antarctic explorer in 1953. He was also accompanied by a Nepalese mountaineer named Tenzing Norgay...And I should also tell you that your son's son also had this distinction of climbing to the top of Mount Everest as well."

"That's fantastic to hear, Dr. Grieg. Thank you for telling me this. But I would have tried to dissuade my grandson from following in my foolhardy steps."

After hearing Mr. Mallory's remarks to what I had informed him, I did not have the heart to tell him that his daughter's husband had been killed in a mountain climbing trip.

Then suddenly I heard Mr. Kennedy ask, "Oh Dr. Grieg, may I speak of my experience now?"

"Why yes. Please do, Mr. Kennedy," I told him.

"Should I start with my angel sighting first? Or how I arrived in here?"

"Start with how you arrived in here."

"Well believe it or not, I was jumping out of a plane, just like Mr. Cooper, Mr. Howard, Mr. Miller, Mr. Saint-Exupery, and Miss Earhart said they were doing before finding themselves in this place. However, this is what is so unbelievable I'm at a loss to explain. The name of the Army Air Corps project responsible for me being on this mission I was on was called - 'PROJECT APHRODITE'!"

"My God!" I loudly exclaimed, as I heard most everyone around me say the same. "That's amazing. But what exactly was this 'Project Aphrodite' you were connected with Mr. Kennedy? And why was it named after this goddess of the Greeks?" I asked, not knowing what to make of this sudden revelation of his.

"Well the purpose of 'Project Aphrodite' was to destroy the V-1 buzz bomb launch sites being used by the Nazis to destroy London. But in order to successfully accomplish the destruction of these V-1 sites without losing any of our men in the process, we had to aim unmanned radio-controlled B-17's loaded with explosives right into them. Of course these B-17's had to first be flown by volunteer pilots and technicians, who later, when their bombers were airborne, were to bail out over the channel. Then unmanned, the radio-controlled B-17's were to be guided by manned mother ships, whose job it was to explode them into these Nazi V-1 missile sites, as I said."

"But Mr. Kennedy, what happened to the men who bailed out of these B-17's? They weren't captured by the Nazis, were they?"

"No. Thank God, Mr. Amundsen, they weren't. They were to be rescued by seaplanes after the B-17's crews parachuted into the English Channel. But I guess I didn't make it, now did I, Dr. Grieg?"

"Well from what I've read about you, Mr. Kennedy, you're mighty lucky to be alive. Because that B-17 you were on was said to have exploded only moments after you and your copilot were supposed to bail out over the channel. And if it hadn't been for the cloud you probably are going to say you jumped into, you'd now be food for the sharks, like maybe your copilot ended up being. But tell me now, why did the Army Air Corps name this project of theirs after the goddess Aphrodite, of all things?"

"Well the project was actually named after a butterfly, Dr. Grieg, the *Aphrodite butterfly*. I guess the army thought, that like a caterpillar, a cocoon stage had to be first gone through before their project could get off the ground. Then again, maybe I'm wrong. Maybe there was another reason for them wanting to name their project after a butterfly, but who really knows the answer to that question?"

Suddenly, everyone heard the voice of His Royal Majesty, Tsar Alexander the First, remark, "Maybe it was because of the belief in Northern Europe that to see a butterfly flying at night was a warning of death, Mr. Kennedy. Plus the fact, that paintings of medieval angels were sometimes depicted with butterfly wings, as were the drawings of storybook fairies. And the widely believed notion that the soul of man has the form of a butterfly. So perhaps your army, Mr. Kennedy, wanted to name their project after what it represented to these Nazis: the angel of death coming after their immortal souls. What better way to symbolize this than to name this project after a goddess and butterfly going by the same name."

"Yes, Your Royal Highness, what you've said makes a lot of sense. And by the way, you sure know your butterflies."

"Yes, Mr. Kennedy," I said, "we can certainly agree on that. But what of the angel you saw? Tell us when and where the sighting of her took place, if you please."

"It was while I was in Rome, Dr. Grieg. Immediately after I had the honor of kissing the ring of His Holiness, the Pope, believe it or not. You see, my father, being Catholic and the U.S. Ambassador to Great Britain had influential connections in Italy."

"You mean you saw an angel in Saint Peter's Basilica, Mr. Kennedy?" we all heard Mr. Hoffa ask, now seeming like Joe Jr.'s friend instead of his foe.

"Yes, Mr. Hoffa, I did. But I thought at the time I

was only imagining the whole affair. I mean look where I was at the time."

"And now you're positive what you saw was real?"

"Yes, Mr. Wallenberg. And must I inform you as to why? But back then seeing an angel meant only one thing to me: that I was soon to be wounded and die."

"But tell us, where in the basilica did you see this angel?"

"I saw her while I was standing before the High Holy Altar, Dr. Grieg, which I knew had beneath it the crypt containing the tomb of Saint Peter himself!"

"Well what was this angel doing there? I mean what did you see her doing?" I said.

"She was praying at the foot of the altar, to the altar's left side. And I could see her face from that angle, Dr. Grieg, and it was the most wondrous face I have ever seen."

"That's all she was doing? Just praying?"

"Not exactly Dr. Grieg."

"What do you mean?" I asked.

"I mean she was also looking at me and saying something. And for a second I thought she was saying my mother's first name of 'Rose'. However, when she began saying my name Joseph many times, I knew I hadn't been wrong about my number coming up because I believed she was praying for me at the time."

"Oh I see. Then what did she do, Mr. Kennedy?"

"You mean after saying my name, Doc?"

"Yes exactly."

"Well you're not going to believe this, but as I followed her into the Sistine Chapel, she began to float directly up and it was then that she disappeared, Dr. Grieg, just vanished out of sight, like Mr. Hoffa said she'd done after his encounter with her," Joe Jr. said, as I watched him look directly up at the cloud above.

"And you've never seen this angel again, Mr. Kennedy?"

"Not that I know of, Doc. Yet isn't that true of everyone, except Miss Earhart?"

"Yes, I believe so. But maybe we'll see her presently. Now I must ask you, Mr. Kennedy, if you can remember anything in your life that might connect you with any of the people here who've told their stories? Or with anyone who has yet to speak, for that matter?"

"Well, besides being a pilot, like Miss Earhart and Mr. Saint-Exupéry, I believe I may have a connection with you, Mr. Howard," everyone heard Joe Jr. say, as he turned from me.

"Oh really? What connection might that be?"

"I was in Spain, Mr. Howard, about the very time you were there, I believe?"

"Hey! I was in Spain as well, Monsieur Kennedy"

"Oh? When was that, Mr. Saint-Exupéry?" Joe Jr. asked with a curious look.

"When I was a foreign correspondent for L'Intransigeant in August of 1936, a month after the Civil War outbreak. And again during the siege of Madrid in the summer of 1937, for the Paris-Soir.

"That's when I was there Mr. Saint-Exupéry. And what about you, Mr. Howard? When exactly were you in Spain?"

"I was in Madrid in May of '43, Mr. Kennedy, as an invited guest of the British Institute. I was invited by them to give a series of lecture's on Bacon's, I mean Shakespeare's *Hamlet*, as well as on my many films after having done the same for Portugal's Teatro Nacional. It was while I was in Madrid, I might add, that I began having nightmares of someone trying to warn me of something. This person I now believe to have been my deceased assistant, Violette, who I was in love with while I was a married man. Violette had died of cerebral meningitis shortly after nursing me back to health from a feverish cold. I know without a doubt now, it had been her who'd been appearing in these nightmares of mine because of the number of séances

I'd been involved in trying to reach her after her death."

"You mean you were also involved with spiritualism, as Mr. Saint-Exupéry and I, Mr. Howard?"

"Yes, very much so, Miss Earhart. I'd become involved in the occult practice of 'automatic writing' and visiting mediums in order to communicate to Violette of how much I missed her. It was the only way for me to get through the anguish I was feeling at the time. You see, I felt responsible for her death. I should have sent for the bacteriologist a friend at the studio had recommended I do. But I didn't believe Violette needed such a specialist for what I considered to have been only a cold. And it's because of all of my guilt and sorrow over her death that I agreed to go abroad to Portugal and Spain. I needed a change of scenery in order to get through this traumatic period of my life, you see."

Turning next to Mr. Kennedy, I asked, "But why were you in Madrid, at about the same time as Mr. Saint-Exupéry?"

"Well I arrived in Barcelona first, Dr. Grieg. The very day it fell to Franco. Then after going to the island of Majorca and boarding a British destroyer bound for Valencia, I joined Woodruff Wallner, the American vice-counsel there. And later the same day we found ourselves unfortunately caught up in a fascist air raid. It was then I decided to get the hell out of Valencia by hopping on a special bus for Madrid.

"However, after arriving there and leaving my duffle bag at the shelled and abandoned U.S. Embassy, I began thinking of leaving Madrid as well. This was because of the bombs that started going off in the city; bombs I knew had been fired from the mortars of Franco's artillery. So was it any wonder then, that Hemingway himself had decided three months before to leave Madrid for his peaceful little home in the Florida Keys, he must have sensed what was coming to Spain very soon. Yet I, on the other hand, only wanted to make contact with Franco's Fifth Column. The reason for this being my deep interest in politics, among other things...So you might say, I was in Madrid, first and foremost, out of just plain curiosity, Doc."

"But Mr. Kennedy weren't you worried your number would come up, like you said? That you might be killed in that civil war going on around you? I mean hadn't that angel you saw make you want to avoid areas of danger from then on?"

"Yes, of course, Mr. Wallenberg. But I figured when my number came up, it didn't really matter where the hell I was. So I gave the thought of my death little concern at that time."

"Tell us how you went about your meeting with Franco's Fifth Column? Just how did you manage the whole affair?"

"Well, Mr. Howard, I was fortunate in meeting a young woman who happened to be the daughter of an American general, she also being the wife of a devoutly Catholic underground leader by the name of Antonio Garrigues y Diaz Canabate. And after meeting them, I was invited to travel with them on their operations throughout the city.

"However, it was only a short time later, Garrigues, and his two men and I, found ourselves stopped at a Communist roadblock in Garrigues little American car, this after being out earlier visiting the prisons trying to get the guards to release everyone. As we knew Franco had won a war about to end.

"Then after being successful in getting these prisoners released, Garrigues, his men and I, were certain we were going to be executed by these Communist irregulars at this roadblock, especially after having been ordered out of our car and lined up against a wall. This happened when Garrigues false Red Cross documentation hadn't worked in getting us on our way. However, because I had presented my American passport and visa immediately afterwards, we were all allowed to proceed unharmed in the direction we'd been going. I suspect these Communists didn't want to be blamed for causing an international incident with an American, in an American car no less."

"You know, Monsieur Kennedy, the same thing happened to me in Madrid. I too was almost executed there."

"Oh really, Mr. Saint-Exupéry?" I said, "Please tell us what happened to you? It might be important. It might be a clue to this crazy puzzle I'm trying to piece together in my mind. So if you please monsieur."

"Oui oui, of course, Docteur. I was in Spain during their Civil War, while I was trying to write a firsthand report of it. Unfortunately, for me though, I had made the serious mistake of doing my snooping around Madrid at the ungodly hour of three in the morning. I had at that time come across the loading of a secret cargo at a freight depot. And because of the darkness and busy activity all around, I was not too concerned I would be discovered in the darkness where I was hiding. But I had underestimated the keen alertness of a small group of anarchist militia who had nevertheless spotted me.

"What happened was this: they had silently approached me undetected, like a den of lions stalking their prey. It was not until I felt a rifle butt make contact with my abdomen that I realized I had been discovered. However, as I swiftly lifted my arms up and into the air, I observed them looking, not at my startled face, as one might have expected, but at my out of the place tie I had stupidly forgotten to remove. It was then I understood why I had become an object of their curiosity. For in the anarchist neighborhood I had ventured on that sad and sorry night, people wearing neckties simply were not to be seen. So during those moments of my most certain death, from what I knew was shortly going to be caused by a single gunshot to my head, I could do nothing but pray to the Lord God Almighty as I waited in my horror to die.

"Yet to my utter amazement, no shot was then fired. After what seemed to be one long and drawn out minute where everyone seemed to be moving around me in a dreamlike fashion, these men suddenly motioned for me to walk on out in front of them.

"Then, after having crossed some nearby railroad tracks, we all entered into a basement of what appeared to be a kind of makeshift guardhouse, where I saw in this dimly lit room about five militiamen seemingly asleep with their rifles leaning up against their spread-opened legs. Upon my entrance into this room, these men began exchanging words with the men who had ushered me in as someone proceeded to search me for any ID and possibly a gun.

"It was then I began to speak to these militiamen in Spanish, hoping they would understand me, I was afraid they all only spoke in their native Catalan. What they seemed to be asking me was where were my papers. Immediately I realized I had forgotten them back at my hotel. So I quickly answered them by saying, 'Hotel...Journalist', hoping they understood me. Yet all they could do was start staring at my camera I had dangling from a strap around my neck. But then I noticed they were not looking at the camera or me a few moments later with too much concern. As I remember it now, they all just seemed to be bored by the whole silly affair, and I did not know why they were all behaving in this unusual manner. I would have preferred a little hostility on their parts to this robot-like behavior they were then displaying. They were just staring at me without reacting, almost as if they were just watching a monotonous chess match. It was as if they were all waiting for something to happen. What on earth that was supposed to be, I had no idea. It was then that I worried over the fact they might do something stupid - like start firing their guns into the air, or at me. Yet was I really in any danger from them? Were all these men really convinced in their hearts I was a saboteur or a spy, and not what I had tried to persuade them: just a foolhardy French journalist?

"Need I tell you, Dr. Grieg, ladies and gentlemen, I was in a cold sweat at that desperate time because I knew these anarchists shot people out of hand and without any qualms. And I was just a man of thirty-seven, too young to die because of a mistake on both of our parts.

"Then a miracle occurred, and a very discreet one at that. You see, I had run out of cigarettes. So noticing one of the men around me smoking, I communicated to him by gesture, if he might give me one, smiling at him as I waited for him to respond. What happened next surprised me. This man began to stretch, pass a hand across his forehead, lift his eyes to my face instead of my tie, and to my utter astonishment, return my friendly smile as if I were his longtime pal. And immediately I felt my fear of this man begin to subside, the atmosphere in this windowless room seemed to suddenly change, as if the dawn outside was filling it with the morning hot rays of the sun. It was as though his smile had set me free all of a sudden, as though his facial expression was seemingly transforming itself from that of indifference to my presence, to that of a fellow compatriot in arms.

"As he continued to smile at me, giving me one of his cigarettes, he seemed to me a little clumsy and shy. I instantly realized his reason: it was because now his human quality showed up his vulnerability. I was no longer considered by him as just a distant object, but as a man with the same feelings as any of them.

"So in response to his kindness I thanked him by placing a hand on his shoulder, which at once seemed to break the ice in the room, as all the militiamen likewise began to smile at me in return..."

"It would seem, Mr. Saint-Exupéry that you, Mr.

Kennedy, and Mr. Howard, have very much in common with each other" I said. "That's our problem however. This common link everyone seems to share. Still, there must be something even more profound about our similar connections we have as yet to discover, besides seeing this angel, as if that wasn't already profound enough."

"Maybe after the people remaining, who have as yet to speak, give their accounts, Dr. Grieg, like you said, we might find out what this deeper connection is, and why we were brought here because of it."

"Yes, Judge Crater, let's hope so. But why don't you be the next one to explain his presence here. Tell us if you please your particular story."

"Well, as you and the others can see, Dr Grieg, I'm a man of forty-one, six feet tall, about one hundred and eighty-five pounds, have mixed gray hair slicked down and thinning at the top, of medium dark complexion, brown eyes, favor a high-collared shirt, and three-piece suit and Panama hat. If you're wondering why I've just described myself to you all, it's because I'm trying to sort of prime my memory for what I'm about to say. Where do I to begin? I just don't know."

"How's about telling us about your angel sighting, Judge, that'll do for starters."

"Well in my particular case, Mr. Dillinger, I'm afraid the circumstances of my heavenly encounter are going to be a might hard to swallow. For you see, I was sort of indirectly involved with this angel most of us have seen. It occurred immediately after one of my brothers, Doug, who had enlisted in the Canadian Air Force, died from shellshock shortly after the First World War. And I wasn't even in my body when this happened to him either!"

"What exactly do you mean by that, Judge Crater?" I asked him with a curious look.

"Well, Dr. Grieg, I had one of those 'out of the

body experiences' people sometimes have. I believe this is also referred to as, 'Astral Projection', or by the East Indian name of 'Eckankar'."

"You mean you actually were able to leave your body and travel to your dying brother, Judge Crater?"

"Yes, exactly, Miss Earhart, and that's when I saw the angel. She was standing at the foot of my brother's hospital bed. And I could smell a strong scent of roses in the air. How I knew my brother was there at that particular hospital, I'll never know. I guess my subconscious mind must have had some inside information about my brother dying. So it must have decided to get me to him immediately by this method of astral projection.

"As to what happened after I arrived and saw this angel at my brother's bed, well I almost died along with him from the shock. For in what must have been only a minute, this being, whose back I'd been facing, with no wings, I might add, suddenly made an about-face and began crying at me. It was then I instantly realized my brother had died. For this being was all aglow with a brilliant white light enveloping her.

"I can't begin to tell you of the compassionate and loving expression on her face, it was a look of such tenderness, I was almost beside myself with a likewise love in return. Then all of a sudden I was back in my body and bed, as if nothing had occurred.

"Nevertheless, I knew I hadn't been dreaming this incredible experience because that very morning I was awakened by the ringing of my phone and informed of my brother's death. When I asked the doctor what the number of the room my brother had died in was, his number matched the one I'd seen when I'd been out of my body in that hospital."

"You're nuts, Judge!" Dillinger suddenly shouted so everyone could hear. "And completely off your rocker. I mean who the hell ever heard of a guy being out of his body and traveling around like a ghost?

Except after you've kicked the fucking bucket perhaps."

"Hey, Dillinger! Didn't I tell you to watch your filthy mouth! You wise ass son of a b…"

"Mr. Hoffa!" I shouted. "Please control yourself. I'm sure Mr. Dillinger didn't mean to say those last few words," I said, as I watched Dillinger give Mr. Hoffa a smiling grin. "So before you kill each other, I'll just remind you, Mr. Howard has my permission to shoot that gun he's holding if necessary."

"Sure, Doc, sure, but gun or no gun, if this two-bit punk dares get out of line again, so help me I'll…"

"OK, Mr. Hoffa, I'll keep that in mind, as I know Mr. Dillinger will," I said, as I saw Dillinger now eying Mr. Hoffa with a murderous look on his face. "But getting back to you, Judge Crater," I went on, "and please forgive our two friends for the interruption. Just exactly how did you get here? I mean what were the circumstances leading up to your strange disappearance?" I asked with concern.

"Well like I mentioned earlier, I'd been involved in a shady transaction of sorts. It had to do with the sale of the Libby Hotel - a twelve-story building. I say 'shady transaction' only because, as the appointed receiver, I had to approve all the bids when the building was finally put up for sale because of a tremendous loss to the hotel's small group of stockholders. And being the steward of the estate, I had to take possession of the funds, and attempt to collect all the monies due.

"So you see, I was quite honor bound to act as the sole treasurer of all this money under my care. When the hotel was eventually sold to the American Mortgage Loan Company, on a bid of a mere $75,000, and then two months later sold by them to the city of New York for the fantastic sum of $2,850,000, because the city wanted the hotel condemned for a street-widening project, the public, including the newspapers across the country, thought those paying this outrageous price were in on the take. Of course this was because of the present reputation of New York's Tammany Hall.

"However, there was absolutely nothing illegal or underhanded anyone could find, in regards to this transaction with the city of New York, or so I thought at the time. It was only when I was shortly thereafter appointed as a judge to the New York State Supreme Court on April 8, 1930, by Governor Franklin Delano Roosevelt, in order to fill the un-expired term of Judge Joseph M. Proskauer, who had unexpectedly resigned, that I began to suspect the Libby Hotel deal had been an inside job. By Judge Proskauer's timely resignation, I couldn't but suspect he knew something of the truth behind the sale. And I knew he just couldn't live with this knowledge and still preside as a Supreme Court Judge.

"You see, he had personal friends he didn't want to rat on who were involved. It was as a newly-appointed Judge of the State Supreme Court, however, that I accidentally discovered who these corrupt friends of Judge Proskauer actually were. All of them it seemed were acquaintances from my recent past. Yet unlike Judge Proskauer, I had no intention of resigning my appointed post because of a personal interest in the matter.

"However, on the evening of August 6, 1930, almost four months after the date of my appointed judgeship, I was abducted by force by men I was about to inform upon. They had been waiting for me outside of a nightclub I was known to occasionally frequent. And it was while we were all in the taxi they had all pushed me into, that we got lost in a thick eerie fog while the taxi was traveling north on the New York State Thruway..."

"You are certain this fog you were in was not a white cloud, Judge Crater?" everyone heard Anastasia ask.

"It might have been, Your Majesty. But the taxi was enveloped so fast by the fog, I really had no time to make that distinction, I'm afraid. Yet from what I've heard told by some of you, I would have to go along with this fog being the inside of some cloud."

"So what happened after your taxi went into it, Judge? I mean was it then you found yourself alone in this hellhole?"

"Yes, Mr. Cooper, only after my abductors decided to stop the taxi and look around. Of course, none of them could see more than a few feet ahead, and this gave me the opportunity I'd been looking for to escape. I just simply began walking away from the taxi and these four desperate men.

"However, as I reached what I knew to be the rocky face of the thruway embankment, I suddenly realized I was being lifted up bodily into the air, along with this foggy mist I was still surrounded by. Then before I knew it, I heard the sound of someone blowing a trombone in the distance. Of course you all know what happened after that."

"Yes, indeed we do, Judge," I said. "But I'm wondering why you alone were taken up, because these other men with you weren't too far from your departure point, as were the men that abducted Mr. Hoffa in his case. It must be because you and Mr. Hoffa saw this angel in your lives, and these other men didn't. So could it be possible this cloud we all seem to have been saved by has some form of a consciousness? A consciousness obeying this angel we've all seen? Or even more incredible: obeying the commands of the Almighty Himself!"

"It would seem so, Dr. Grieg," Judge Crater was saying. "And it would also seem that this pillar-shaped cloud above our heads may be the Lord's heavenly command post. Or what the Holy Bible calls: 'The Cloud of the Eternal', the very cloud that followed the ancient Israelites in the desert five thousand years ago!"

Chapter Nine

It was only a few seconds after Judge Crater had told everyone of his belief the strange cloud above us was what I believed it to be, that I began to give up the idea we were all inside of some giant UFO. Judge Crater's story of his angelic encounter and abduction by a cloud had been all I needed to hear to convince me of this fact. And I now knew without a doubt we were all dealing with something supernatural, if not the supernatural Lord God Himself.

However, there were still six other people to relate their experiences, and I wondered to myself if they would reveal anything in their stories that would change my mind. But what difference would it make anyway for we were still prisoners in some timeless domain?

Then again, there was a great difference - UFO beings might kill us, or have all of us for breakfast after experimenting on our bodies, but never would the Almighty Lord. He would never bring harm to us. Hadn't all of us been saved by whatever had brought us here?

All the same, if it had been God Himself who had been responsible for all of our fates, then why had He commanded one of His angels to bring us to this place He alone knew? If in fact it had been an angel that had brought us here and not just a mysterious cloud. I could only wonder at the time. Perhaps as I questioned the others, the perplexing answer to this mystery of all mysteries would be revealed. Or perhaps our 'guardian angel' friend the people I had questioned had seen would come to our rescue again, telling us not only why we had been chosen to be here, but also returning us to our individual times.

So hoping again for some miracle of a clue to

our way out of here, I turned to Mr. Rockefeller and said, "Mr. Rockefeller, why don't you tell us your story now. When did you see the angel? And why were you, in of all places, New Guinea, being chased by these wild cannibals out to kill you, only moments before arriving here?"

"I was in New Guinea, Dr. Grieg, taking part in an expedition sponsored by Harvard University's Peabody Museum. I'd signed on as a sound technician. The purpose of the expedition was to film and record the customs of New Guinea's little known tribes.

"Within days of being in New Guinea, however, I began taking photographs myself and became so fascinated by the artifacts the natives made, I personally started collecting as many of them as I could find.

"You see, I wanted to give these artifacts to the museum of Primitive Art in New York, which had been founded by my father, and of which I'd been named a trustee.

"As to why I was being chased by headhunters trying to kill me, well that will take a little bit of explaining. So do you really want to hear about this, Dr. Grieg?"

"Yes, of course I do, because the events leading up to your sudden arrival here may give us a clue as to why you're still alive, or should I say why you were chosen to be saved from a certain death by a cloud."

"Yes, that cloud is quite a mystery, isn't it?"

"Yes, it is. And I think I'm beginning to fit the pieces together as to its origin. So tell us, Mr. Rockefeller, what were you doing that day before being hunted down like a dog?" I asked.

"I was in the Asmat coastal area of New Guinea looking for more artifacts for the Museum of Primitive Art.

"You see, after the Peabody Expedition had officially ended, I returned to this area after having gone to

America for a week because of an announcement I heard about my father wanting to divorce my mother. I thought I could persuade him to change his mind in the matter, but it was to no avail. So after telling him I wanted to become an anthropologist, I met a Dutch anthropologist by the name of René Wassing back in New Guinea, where we began exploring the rivers of the Asmat in our motorized catamaran.

"However, it was after we had collected many wonderful artifacts that we decided to save some time by not going to another village by way of the inland waterways. So we instead began venturing into the coastal waters we'd been warned not to enter in our small overloaded craft.

"We were going to enter the South Eilander River, but before we could enter it a large wave swamped our catamaran and turned it upside down. René Wassing and I then managed to hold on to the overturned craft throughout the night as it carried us out to sea because of the outgoing tide.

"There were also two other men who came along with us on our trip. They were two local natives, and before night arrived, they had decided to try the long eleven-mile swim to shore in order to get us some help. We had tried to dissuade them from going because of the many sharks and crocodiles we knew to be in these waters, but they couldn't be argued out of their swim. By morning, with no help in sight, I decided to do as our two natives had done and make the eleven-mile swim to shore..."

"Monsieur Rockefeller, how could anyone have made a swim of eleven miles without tiring out and drowning, as your two native friends must have done?"

"They had both tied on some of our empty fuel tanks to their shoulders as floatation devices, Mr. Saint-Exupéry, as I also did before leaving the catamaran the next day..."

"Then what happened to you, Mr. Rockefeller?"

we all heard Mr. Amundsen ask. "I mean when you finally reached the shore? And why were those headhunters chasing you right afterwards? Or didn't you know that some of the natives in the area of your expedition were certain to be hostile?"

"Well, Mr. Amundsen, I must tell you that in New Guinea, any of the natives can turn hostile if anyone in their tribe is killed, be it because of a white man, or otherwise.

"You see, they have a custom in New Guinea called 'Payback', and when I reached the shore after my long swim, I knew I was probably in the most dangerous and mean populated area in all of the country, as far as the custom of 'Payback' was concerned. I knew this because I knew I was in the vicinity, if not the actual village of the Otsjanep headhunter warriors. When they saw me come out of the water, I knew from their mean expressions, one of their people must have been recently killed because they immediately began to chase me, throwing their poison tipped spears at me in the process..."

"You mean they wanted to kill you in order to avenge the death of one of their comrades?"

"Precisely, Your Royal Highness. You see, the natives of New Guinea believe the spirit of anyone murdered will forever wander the earth if the custom of 'Payback' is not exacted, all the time appealing to its relatives to avenge him so he can go to his final rest. So can you really blame those warriors for chasing me? They saw their opportunity to avenge one of their people and took it."

I then said to Mr. Rockefeller, "But let me ask you now, in your opinion, what do you think may be the reason for you being here? I mean what do you think you may have done in your life you feel guilty about? And please tell us of your angel sighting as well, Mr. Rockefeller."

"I may be here because of some of the things

my great-grandfather, John D. Rockefeller, did in his business dealings during his life. Perhaps I'm being made to pay for them now."

"Like what things?" everyone heard Miss Earhart ask with a sympathetic voice.

"Like for example, the condoning of great-grandfather in the shooting down of innocent strikers trying to form a union at a mine he partly owned in Ludlow, Colorado, in 1913. There were forty people killed by the company mine guards as a result of that strike, including two women and eleven children who suffocated in a cellar when a tent above it burned down..."

Suddenly we all heard Mr. Hoffa remark, "If I'd been there with those strikers, kid, all the dead would have been on your great-grandfather's side."

"Yes, Mr. Hoffa, you probably would have had those miners organized into a union in no time at all. However, my great-grandfather could never live down the killing of all these people, and he was so hated after this by the public that the pastor of his church even employed armed Pinkertons to mix in with the people in the pews, just in case someone was present in the church to do great-grandfather harm. It was because of his bad reputation in his business dealings that he decided to become a philanthropist to his fellow man."

Suddenly we all heard His Royal Highness remark, "Charity shall cover a multitude of sins. So says the Holy Bible, Mr. Rockefeller."

"Yes, Your Royal Highness. However, my great-grandfather always believed in giving part of what he earned to help others in need. This he even did before he'd become a wealthy man. Of course, as President Roosevelt once remarked, 'No amount of charities in spending such fortunes can compensate in any way for the misconduct in acquiring them.'"

"What about 'Attica', kid? Oh, I forgot. You

wouldn't know about that. Attica was after your disappearance."

"What's 'Attica', Mr. Hoffa?" Mr. Rockefeller asked.

"Oh, that happened when your father was the Governor of New York, when he refused to come to Attica prison during an uprising, thereby causing ten hostages and thirty-three inmates to be killed by the National Guard in 1971."

"Well maybe that's another reason why I'm here, Mr. Hoffa. And like you said, that incident was after my disappearance, and I can't for the life of me really accept what my father did as a consequence to why I'm now in this situation. And as far as any connections my family or I have in common with anyone here is concerned, well my grandfather, John D. Rockefeller the Second, helped finance the Admiral Byrd polar expedition, which relates to Mr. Amundsen's polar expedition. And my father, Nelson, along with his brother, Laurence, even went on an expedition of their own to Antarctica with Sir Wilfred Grenfell some years later. And then there's the connection my father's brother, David, had with your family, Mr. Kennedy."

"Oh? And what was that?"

"Well it seems, Mr. Kennedy, that my uncle David, when he was in London trying to enroll in the School of Economics, was invited to a cocktail party where he met your father, Ambassador Joseph P. Kennedy. And then later, my uncle David even took your sister, Kathleen, out on a date."

Suddenly everyone heard Bormann say, "What about the connection your grandfather had with my country and the Nazi party in the 1930's, Mr. Rockefeller?"

"I told you to shut your dirty mouth until your turn came, Herr Bormann!" we all heard Mr. Hoffa then shout.

"That's all right, Mr. Hoffa," I said. "Let Herr

Bormann have his say. He may have something important to add this time."

"Danke, Herr Grieg. I just wanted to bring to Herr Rockefeller's attention, the fact that his grandfather's oil company, Standard of New Jersey, had been doing business with the German petrochemical monopoly, I.G. Farben, since 1920; then renewing his agreement with this Nazi cartel in Holland in 1939."

"Yes, I know all about it, Herr Bormann. Nevertheless, my grandfather had no idea I.G. Farben had been using slave labor from your Nazi concentration camps. But this was before World War Two broke out anyway."

"True, Mr. Rockefeller," Mr. Hoffa was saying. "Still, what about your uncle David's bank, the Chase Manhattan, doing business with the country of South Africa?"

"Well, my uncle David was totally against 'Apartheid'. And he only wanted to help South Africa build and develop itself by opening a branch office there in 1959, then offering South Africa $40 million in revolving credit for that purpose.

"However, I saw his dealings with South Africa in a different light. This was a whole year after the United Nations Assembly had voted to condemn the racial policies of this country. By dealing with their government, my uncle's bank only made it more possible for South Africa to buy even more guns, thereby keeping the blacks there under even more white government control..."

"Mr. Rockefeller," I said, "did you know your father became the Vice-President of the United States in 1974?"

"Boy, that's a real surprise. Even so, my father was always in politics in one way or another during his career. But let me tell you now of the angel I saw as a child, Dr. Grieg."

"Yes, please do," I said.

"It was right after I first saw a statue of the goddess Aphrodite at 'Pocantico', my great-grandfather's mansion in upstate New York. The statue was a copy by Praxiteles, the famous Greek sculptor, who lived in the second half of the fourth century. The only other copies of this statue I know, exists today in the Vatican in Rome, and the Metropolitan Museum in New York."

It was then that I realized the connection the angel sighting Joe Kennedy Jr. had seen in the Vatican and this statue of Aphrodite of Mr. Rockefeller's. What was I really dealing with here? An angel? Or a goddess? Only time would tell, I surmised.

"Well anyway," everyone heard Mr. Rockefeller continue, "it was right after I saw this statue of Aphrodite that I began to fall into a kind of trance in the garden in which this statue was displayed. I was about ten years old at the time."

"Were there any roses in that garden, Mr. Rockefeller?"

"Yes, many, Mr. Howard. Do you think that may have had something to do with me seeing an angel near this statue only moments after I smelled a rose?"

"Well, the goddess Aphrodite had many followers, and still does. She had many cults. One cult worshiped her as the goddess of the sea, another as the goddess of the underworld, another for spiritual love, one for sensual love, and lastly, as the goddess of the people and war ... And as I remember it now, the rose was holy to her, as were the swan, the dove, the sparrow, the poppy, and myrtle. She was only given offerings of flowers and incense as well, I believe....And oh yes, she was also the daughter of the Greek god Zeus, the chief ruler of all the gods and men."

"Mr. Howard, how do you know all this about the goddess Aphrodite?"

"Oh, I studied Greek mythology and drama in my youth, Miss Earhart."

"You remind me, Mr. Howard," His Royal Highness was saying, "of what Michael Maier, the German alchemist, said in 1618. And it was: 'He who would attempt to penetrate into the Rose Garden of the Philosophers without a key resembles a man who would walk without feet.' And the key to getting us out of here I think is about to arrive, if I am not mistaken."

Suddenly, at the finish of His Royal Highness' interesting comment, we all heard in the fog, not more than a few yards from my abandoned vehicle, a great shuddering sound of vibrating metal. Only moments later when the fog from which this sound was coming began lifting skyward, as though caused by a gust of wind, I instantly realized what had made this frightening ululation: it had been the metallic reverberation of the backup vehicles materializing.

"My God!" I shouted. "It's the other vehicles, Professor Moebius, and Elaine!"

On hearing my loud and excited voice, the professor and Elaine, obviously overjoyed in finding me, began yelling out my name, Elaine also saying, as she came running toward me, "Thank God we found you! We thought for a moment you were an illusion. Are you all right, Philip? And where for heaven's sake are we now?"

"Yes, my boy, just where the devil is this place we and these people are in?" the professor asked.

"That's what everyone here has been trying to figure out for the last few hours, Professor. If hours is the correct measure of time to use in this place?" I said. "And am I glad to see you both. But do you realize what you've gotten yourselves into by coming here?"

"There's no need to tell us about that, Philip," the professor was saying, "or what you and all these people have been talking about. Because believe it or not, Dr. Rogers and I have heard just about everything that's been said since you disappeared from the shell."

"But how is that possible, Professor?" I asked,

somewhat confused.

"Yes! How can that be?" everyone heard Mr. Howard remark.

"Well Philip, immediately after you vanished with the vehicle, Dr. Rogers and I frantically raced into the shell to find out if you were still there. And the moment we entered the place, we were amazed to discover, besides the fact you were nowhere to be found, that your voice was echoing off of the shell's walls. It was as though the shell were acting like some kind of a giant transmitter, amplifying your voice from wherever it was coming. At first I didn't believe it was your voice I was hearing, I thought it was the voice of the being in the shell I'd earlier heard. However, when Dr. Rogers remarked to me she was also hearing your voice, I knew I wasn't being influenced by my imagination.

"And let me say, Philip, Dr. Rogers and I are here because of your suggestion to follow you in the backup vehicles with the needed tracks to construct another strip. Of course we would have come after you anyway. Dr. Rogers is almost certain from what you and the others have discussed that there may indeed be a number of ways out of here and back to our world and time. And that includes getting back with or without the strip and vehicle. My idea about reversing the direction of the magnetic field surrounding the shell, as well as putting a full twist into the strip to get you back, unfortunately failed to work I'm afraid."

"You see," Elaine said to me, "when Professor Moebius and I took it upon ourselves to come find you, we realized why this angelic being you and the others have spoken about chose to make her appearance in the area of Corpus Christi, Texas, if she was indeed an angel that we actually saw."

"You mean you saw the angel in the shell Elaine?" I asked with some fear.

"Well, yes. Still, I can't say that for sure, Philip, even though both Professor Moebius and I saw her.

Nevertheless, the name 'Corpus Christi' means: 'Body of Christ'. And whoever named the city of Corpus Christi, not to mention, Padre Island, may have done so for reasons of a holy nature we're now unable to explain. Also, there's the fact of Professor Moebius' ancestors being related to Martin Luther, the religious reformer."

"Then what do either of you have to say about this place we're marooned in? You both must have some theories about the matter after what you've already heard these people relate?"

"Yes, we do, Philip," the professor was saying. "However, I'd rather tell you our theories about this place after Dr. Rogers and I have heard everyone speak, like you yourself had planned. As Sherlock Holmes or Mike Hammer once said, 'One should throw out every farfetched explanation, and the remaining one, no matter how absurd, must then be correct.' And besides, we don't want to prejudice you with our theories as yet. We'd like you to do as you were doing and see where it leads us. As for now, I think constructing the strip would be more than a good idea, dear boy."

"Yes, I'm forgetting you brought those metal tracks with you. And God help us all if after we've put the strip together, the vehicle doesn't function as planned. Because there may not be enough of a magnetic current of force operating beneath this strange metallic floor of silver we're all standing upon to make the vehicle behave like it did back in the shell."

"That's very true, Philip. But don't let that bother you because I don't think we're going to need such a magnetic area to get the vehicle to dematerialize back to our world. I think we may only have to get the vehicle to vibrate out of phase with our surroundings. This may either move us forward or backward in dimensional time. I can't be certain of which, or if it even matters. However, Dr. Rogers has another theory she should tell you before we assemble the strip."

"Yes, I do. You see, I did some reading after your disappearance, and I discovered a number of interesting things...."

"Like what?"

"Well, there seems to exist down through history a connection between sacred geometry and sacred geography, from what I've read; a connection that has to do with secret doorways opening onto unknown aspects of the earth when points of telluric current intersect, beneath either a sacred church, or area where your geometric temple of a shell is located. Of course there's more to it than that. And that's the part about the angel and the reason behind all of these peoples' disappearances."

"Well what's your theory about this? I'd like to hear it now. It could tell us where to position the strip perhaps."

"OK, Philip. Now as Professor Moebius explained to me, our entire universe seems to be constructed according to systems of harmonic resonance, or vibrational harmonies. So what this means is if we can somehow make contact with this angel, whoever she might be, we may then be able to convince her to return us back to our world, as you previously mentioned. How harmonic resonance comes into this has to do with you, or one of the other people here acting as sort of a medium. What I mean is one of us has to make a vibrational mental contact with this being. And it would seem Miss Earhart, Mr. Saint-Exupéry, Mr. Howard, and even you, Philip, have had some experience with mental telepathy of some sort. Your mental reaction with Whitman's poem, *Leaves of Grass*, you had earlier, may be the very thing that might put you into the right state of mind we're going to need to make this contact with her. Plus the fact you've already heard her voice. Because if her origin is of a cosmic nature, then perhaps it can only be contacted by a mind in such a state as its own: a 'cosmic conscious state' I mean."

"That's probably true, Elaine," I said. "But first let's try getting out of here with the vehicles, because there's no telling what's going to happen if we summon her. She might get mad and send us straight to hell, if we're not already there, in fact."

"Yes, you may be right," the professor commented to me. "But even if we should all get out of here safely by means of the vehicles, will this being come after us? Because we don't even know why we're here. She might very well decide to bring us back to this place by means of that cloud above."

"That's a chance we'll have to take, Professor. But let's get the hell out of here first," I said, as I walked over to the backup vehicles and began removing the long, curved metal tracks of the strip, wondering to myself how the backup vehicles had all materialized in this void at the same time, after the professor had told me they had been sent after his, unmanned, one at a time and twenty minutes apart.

Chapter Ten

Two hours later, it seemed, after having introduced the professor and Elaine to everyone, some of the men and I managed to put together all of the interlocking parallel tracks of the strip, including the half twist in two of the curved tracks, after which we carefully lifted my lesser weighing and shorter vehicle and its wheels on to the strip by way of a ramp.

The entire Moebius strip had to be off of the ground by at least four feet, so the vehicle could clear the ground when it turned the twist and rode the strip's outside surface. This we had accomplished by putting one four foot T-shaped metal strut at each of the strip's lower four corners where the tracks arched up - the lower part of the strip being the part without the twist. This had also been done by my assistants back in the shell. The parallel tracks of the strip were themselves held together with thin metal struts almost exactly like regular railroad tracks, which used wider wooden ones instead.

All of the metal tracks of the strip had been made with an interlocking feature we had incorporated into them. The metal tracks were made so that they could be fitted together like the tracks of a toy train. The backup vehicles had been constructed to haul this extra load because the professor knew the strip would not be able to enter the void we were in when my vehicle materialized there. And this fact had been confirmed when I had vanished from the shell without the strip.

Then having to decide if I was going to leave Elaine and the professor, as well as all the others alone, while I was riding my vehicle back to who knew where, I turned to Elaine and said, "Elaine, do you realize if I should happen to get myself lost again in some other dimension by this second ride, I might never see you again? And I mean never?"

"Yes, I know, Philip. But what choice do we have, other than the mental experiment I told you about. So maybe we should try it instead of your ride now."

"No, I don't think it would be a good idea at the moment, because we might summon something other than this being in here. There are just too many unknowns to worry about. So I guess it's goodbye again, Elaine," I said, as I gently took her hand into mine and kissed her on the lips. Then turning to the vehicle without a word, I got in, closed its glass canopy over my head, and started the engine now fully fueled.

As I sadly waved to Elaine, the professor, and all the others, Bormann not included, I feared for a moment I would undergo the unbearable pain I had experienced when I had first entered this world.

Still, due to the fact neither Elaine nor the professor had mentioned what I had experienced, I put this thought quickly out of my mind.

It was only a minute or so later, however, when the vehicle had made about six trips around the strip that I began to quickly realize the vehicle wasn't going to dematerialize like we were all praying. There seemed to be a complete lack of whatever it had been that had caused the vehicle to react in the shell like it had back at Port Aransas. Perhaps it was the shell itself that was the needed requirement for the vehicle to return to the time I knew, I wondered. So maybe what was needed was another shell for my roundtrip journey? However, that shell I had found washed up on the beach as a boy had to have been one shell in a billion. So finding the likes of another shell of its size in a place devoid of even a pool of water, or a grain of sand, was an impossibility, to say the least.

As the vehicle made what I estimated to be the twenty-third trip around the strip with nothing at all happening, I sadly lifted my foot off of the gas pedal and turned off the engine. After exiting the vehicle with an expression of utter hopelessness on my face, I said in a monotone of a voice, "Well that does that, Professor. Now Elaine and you are trapped in here as well, I'm afraid."

"Don't give up so soon, Philip. Remember, I told you Dr. Rogers and I had a number of theories about getting us back home. And you already know about Dr. Rogers' idea of holding a séance. All the same, before I tell you my other personal theories, or have Dr. Rogers hold this séance with you acting as the medium, I think you should continue to question the famous people here. I believe there are five more of them who haven't as yet told you of their angelic encounters? Or under what circumstances they arrived here, am I correct?"

"Yes, you are, Professor. I think I'll ask Mr. Glenn Miller to tell us his story next...."

So turning to Mr. Miller I said, "Mr. Miller, I'd like to question you now, if I may? Can you tell everyone about your angel sighting please? We already know how you arrived in here. I believe you said your plane's wings iced up in a heavy fog over the English Channel, thereby forcing your plane to crash?"

"Yes, Dr. Grieg, that's what happened all right. I should have realized we weren't going to make the crossing in that fog. Thank God for that cloud I was blown into though, or I wouldn't be alive right now because we had to fly very close to the surface of the channel's choppy sea. The plane didn't have deicing equipment and we'd been ordered not to send an SOS under any circumstances. You see, the 'Battle of the Bulge' was about to start."

"Well then, just why were you flying across the English Channel, Mr. Miller? Why was it so important you were willing to risk your life flying in that fog?"

everyone heard Miss Earhart ask.

"I wanted to get to Paris ahead of my band to find accommodations for them, Miss Earhart. You see, I had the day before already postponed the trip because of the weather. Two days before, as a matter of fact. So when I was offered a ride by Flight Officer Johnny Morgan in General Goodrich's personal single engine nine-seat C-64 Norseman because of the general being sick at the time, I couldn't pass up the opportunity. You see, I knew there were no German fighters in the area, or enemy artillery within two hundred miles."

"But what of your angel sighting, when was it and where?

"Dr. Grieg, you aren't going to believe this, because I saw this kind of ghostly figure appear right after I began playing a tune called *And the Angels Sing.* The tune's based on ancient Hebrew themes and tempos, and was authored by Ziggy Elman. The lyrics were written by Johnny Mercer, and guess where I first played it which caused this being to materialize, Doc?"

"Where, Mr. Miller?" I asked, knowing I was in for another surprise.

"At Boston's *Roseland-State Ballroom*!"

"Did anyone else see this being, besides yourself, when it appeared to you?"

"No. I was alone, Mr. Howard. I was at the *Roseland-State Ballroom* an hour before any of my band arrived, or anyone else for that matter. Of course, the ballroom manager was there to let me in, but he was in his office and not in the ballroom until an hour later."

"Well let me ask you this then: were there any roses in the ballroom that you noticed?"

"No. Not a one, Doc. But now that you've mentioned it, I did smell roses ten seconds after playing that tune, I believe. And when I stopped playing it the smell of roses seemed to quickly fade. I didn't know what to make of this at the time. Nevertheless, let me tell you what I saw as I was playing *And the Angels Sing*, Dr. Grieg."

"Yes, please do."

"Well as I began playing this tune, I almost choked on my spit when I noticed a strange ghostly figure begin to take form in the middle of the ballroom floor. Of course, when this started I immediately stopped playing. And when I did, I noticed the apparition begin to fade almost at once. So I began playing the tune again figuring I had only seen an illusion of sorts. You see, Doc, I'd been known to hit the sauce a lot, so I knew from experience I might start seeing anything from pink elephants to green fairies. Yet when I started playing that tune again, I saw the same ghostly figure materializing in the center of the ballroom. Only this time, Doc, I knew it wasn't an illusion, because the thing started to get brighter and brighter until it took on the form of the most beautiful woman I have ever seen, and I almost passed out from her glorious sight.

"It was then I quickly realized she was no ordinary woman I was seeing, but an honest to God angel. I knew this because she wasn't only floating off the surface of the floor, but was dressed entirely in glowing white clothes."

"Then what did she do?"

"She began to pray, Doc. And it was then that I stopped playing again. Only this time for good. And when the guys from my band showed up an hour later, it took them almost an hour to convince me to rehearse the same tune with them. I only did so because I figured what I'd seen wouldn't happen a third time with so many people around. I knew the angel had appeared for my eyes alone. As to why she had appeared, Doc, that I could never solve. Not until now that is."

"I'm wondering, Mr. Miller," I said, "what would happen if you started to play that tune again? Would this angel appear to us perhaps? I mean isn't that what we'd be hoping to achieve by the séance we were going to hold?"

'Well, what does the professor and Dr. Rogers have to say about this?" everyone heard Mr. Miller ask.

"It sounds like a good idea to me, Philip," the professor replied to me.

"With me also," Elaine agreed.

"OK, Mr. Miller, I guess you have the green light to proceed. But aim your trombone at that cloud above. Because I've got the feeling this angel of ours is up there just watching us all, if not the Lord Himself."

As we all heard Mr. Miller blow the first note of the tune *And the Angels Sing*, I began reciting out loud the *Ninety-First Psalm of David* to myself. And as I did so, I heard, to my surprise, everyone, with the exception of Bormann and Dillinger, reciting the psalm as well.

"He that dwelleth in the secret place of the most High shall abide under the shadow of the Almighty. I will say of the Lord, He is my refuge and my fortress: my God; in him will I trust. Surely he shall deliver thee from the snare of the fowler and…"

Suddenly, from the cloud came a crackling bolt of lightening, followed almost instantly by the clamorous sound of thunder, and I knew something supernatural was about to appear.

"My God! Dr. Grieg! What's that coming out from the cloud?" Mr. Howard began shouting to me.

"It looks like it's the angel! But she seems to be pointing or something," Miss Earhart was saying, somewhat dismayed.

"Oui, Dr Grieg, the angel is pointing! Just like in the Sahara desert with Prevot and I! And she looks like the same angel!" Mr. Saint-Exupéry was frantically shouting.

"Well maybe she's trying to tell us the way out of here, Dr. Grieg!"

"Yes, Mr. Mallory, I believe you're right. She seems to be pointing in the direction to go."

"Yes, but which way is that? We don't even have a damn compass. Or do we?" Mr. Kennedy was asking me.

"Jesus! I've got one! Will it work in this place though? Let me see," Mr. Amundsen said, as everyone watched him reach into his furry parka and pull out his compass. "Damn it! There seems to be a magnetic fault somewhere. Just look at the way the needle is spinning," he finished saying, as we all gathered around him and curiously watched his compass's strange behavior.

Then as we all looked back up at the angel hovering like a holy hummingbird beneath the cloud, we saw a glowing white beam of light, similar to a laser, begin to emanate from the finger she still had pointing.

"Well I'll be. Just look at that. She's got a damn flashlight in her finger, Doc!"

"Not exactly, Mr. Dillinger," I said. "But whatever that is, it'll keep us from getting lost. That light's better than any compass because all we have to do is follow it. It'll be our lifeline out of here. At least that's what I'm praying. Unfortunately, if there's anyone else lost in this fog, they're also going to notice that beam of light. And wherever it leads us, will most likely lead them as well. So let's hope we don't run into anyone before we reach the exit to this place because there may still be dangerous people around."

"In that case, Doc, maybe you should start passing out the guns Mr. Howard has."

"OK, Mr. Hoffa. But don't give Dillinger one."

"Do chickens have lips, Doc? Me give that nut a gun? What do you take me for?" Mr. Hoffa laughed.

"Does that include me, Herr Grieg?" Bormann asked, with a grin as big as his gall.

"Shut up, you Nazi tub of lard! If I'm not getting a gun, you sure as hell ain't getting one either!" Dillinger shouted in a rage, as I watched Mr. Howard give all the men their pistols back - Mr. Howard keeping Dillinger's, and I keeping the Luger of fat Herr Bormann.

Chapter Eleven

It was about a few hundred feet into our walk following the light beam, when Mr. Miller turned to me and said, "You know, Doc, this fog is nothing like the fog I saw when I was in London in December of '44. And just to give you an idea of how thick it was, you know those double-decker buses London is famous for? Well, the conductors had to walk three feet ahead of their buses with a flashlight pointed back toward the drivers so they wouldn't run their buses over the curbs or crash them into some other vehicles.

"Then there was the time when two of my friends and I had to crawl on our hands and knees back to our hotel after coming out of a restaurant we'd dined in. We had to feel the sides of the damn curbs for a whole lousy mile. Now how's that for fog, Doc?"

"Yes, Mr. Miller, I guess we're pretty lucky we can see a hundred feet ahead. However, let me ask you a question I failed to ask you."

"Oh, you mean about any connections I might have with any of the people here?"

"Yes. That's right," I said.

"Well I think I might have a connection with Judge Crater. You see, before I joined the Army Air Corps, I had a young booking agent working for me whose father held the post of Supreme Court Justice for Westchester County in New York, if that means anything?"

"Oh, Herr Grieg, I seem to have a connection with Herr Miller. May I make mention of it now?"

"Yes, Herr Bormann, you might as well. What is it?" I asked him with disgust.

"Well my father played the trumpet. He was a

trumpeter-sergeant-major while in the army. And Herr Miller is he not also a major in the army and a player of the horn?"

"Yes, but..."

Before I could finish my reply to Bormann, I saw emerging from the fog a hundred feet to my left, what looked like a whole squad of Nazi soldiers dressed in the black uniforms of the dreaded SS.

"My God!" I shouted "There's Nazis here! Look! Over there!" I shouted again, now pointing their way.

"Looks like my bodyguard found me, Herr Grieg. And I think it would be a good idea if you tell everyone with a gun to hand them over to me now. Then I would suggest you and everyone to put their hands up. Unless of course you want my men to shoot you all down like the dogs you are."

"Yes, Herr Bormann. I can see we're outnumbered," I said, scared to the bone.

"Hey, Doc, let's put a gun to this kraut's head. I don't think his men will shoot anybody then."

"No, Mr. Hoffa. I don't think it's going to be necessary. Because I believe I can see a whole battalion of British soldiers directly behind them," I said, now quite relieved, yet also amazed.

"You're bloody right, old man. There is a battalion of British soldiers over there. And I think it would be a good idea if we all began to duck because there's going to be a whole lot of bullets flying our way. We seem to be in their line of fire," Mr. Howard warned me.

"Those British soldiers, Mr. Howard, must be some of the men you mentioned disappeared during the Battle of Gallipoli during the First World War."

"Yes, I think they are. But will they shoot at those Germans? Because those German uniforms of theirs don't look a thing like the uniforms the German Army was wearing in 1915?"

"Don't let that worry you, Mr. Howard, because when those krauts see who's behind them, they'll start shooting regardless," Mr. Kennedy said.

"Then maybe we should all lie down on the ground as Mr. Howard suggested."

"Yes, Judge Crater, I think that would be a good idea right now." I said, as I told everyone to spread-eagle on the foggy floor.

It was no more than fifteen seconds later that we all heard the sound of many guns being fired. And another fifteen seconds when all the guns became suddenly silent again.

"Hey! It sounds like the shootings over, Doc, but who won the war?"

"I don't know, Mr. Cooper. All the soldiers seem to have disappeared. I wonder if that's also true of the dead soldiers laying on the ground and hidden by the fog?"

"I'm not sticking around to find out. They may only be wounded, which means they could start shooting again the moment we walk their way."

"Perhaps you're right, Mr. Dillinger, so we all better follow that beam of light like we were doing. And as for you, Herr Bormann," I said, facing him, "you can lead the way, just in case any of your men survived and we run into them again," I finished telling him in a stern and vindictive voice, as I next turned to Mr. Howard and said, "Mr. Howard, you think we should get the men to give us their guns back now? Or should we let them hold on to them?"

"I think it would be a good idea if we let them keep their guns, old chap, because there's no telling who we're going to run into following that light beam. Nevertheless, you keep an eye on Mr. Hoffa and Mr. Kennedy, and I'll keep an eye on Bormann's bleeding ass, which I just might use as target practice, by Jove."

"That sounds like fun, and I guess you're right about the guns. However what if…?"

"Hey, Doc!" I suddenly heard Mr. Hoffa shout. "Do you smell the odor of rotting eggs? Or am I the only one in here smelling that?"

"Hey! Hoffa's right!" Dillinger yelled. "I also smell rotting eggs."

"Good God, Dr. Grieg!" Mr. Howard was saying. "But doesn't sulfur also smell just like rotting eggs?"

"Yes, he's right, Philip," Elaine exclaimed. "Sulfur smells exactly like rotten eggs. But what seems to be the matter? You look worried all of a sudden?"

"Yes, Philip, what seems to be the matter? Everyone's looking very scared."

"I explained earlier to them, Professor, that the smell of sulfur where there isn't any around may indicate an evil presence in close proximity."

"Oh yes! I did hear you speak of that while Dr. Rogers and I were in the shell. However, there could be another possibility to explain this smell of sulfur, one that might very well explain why there's a fog in here and why there's a silver floor under our feet."

"You don't mean that we're all inside the crater of a volcano, Professor?"

"Yes, precisely! This silver floor must be the result of silver ore deposits melting into a smooth surface and then re-solidifying again when the volcano from which they originated became dormant. The electrical sparks under our feet can be explained as well: they're probably nothing more than a buildup of static electricity from the atmosphere, because we're most likely very close to an elevation of twenty thousand feet. And the fog all about is probably due to the volcano we're standing upon becoming active again and venting steam, thus the odor of sulfur from the volcano's hot core."

"My God!" I said. "But that would mean there may be a way out of here. That is if we can find the rim of this crater we're in. And God help us if we don't. Because if this is a volcano at the top of which we're standing, a volcano that's become active, then we may all be fried to a crisp within hours. Unless of course that angel we saw saves all of our lives again."

"Well at least we have an explorer with us who discovered the South Pole, as well as a famous mountain climber."

"Oh, you mean Mr. Amundsen and Mr. Mallory, Professor," Elaine said. "But what happens when we reach this crater's rim and it's too high to climb? Because we don't even have a rope or any climbing equipment."

"We'll worry about that later when we see what we're up against, which I might add, is directly ahead, I believe I can faintly make out this volcano's rim, if I'm not mistaken."

"Great Caesar's ghost! You're right, Professor! I can see it too!" everyone heard Mr. Howard shout.

Turning to Mr. Mallory I said, "Mr. Mallory, do you think you can scale that rim? It looks a mile high."

"I can't tell. It's too soon to say, Dr. Grieg. But from the looks of it, I don't stand a chance without any climbing equipment. I left all my gear back at my base camp on Mount Everest. But who knows, there may yet be another way to scale that rim. Or should I say through the rim? Because there might very well be an opening, or cave, leading to the rim's other side."

It was then I knew I was going to have to hold that séance Elaine had talked about earlier. It now seemed the only way we were going to get out of where we were, was to contact our angel and plead our sorry case to her, hoping right afterwards, she had it in her power to do something about saving us.

Still, why had she pointed the way to this crater's rim knowing we had no chance of climbing it? Had she only wanted to tell us we were inside of a volcano? And if she had, then what for I wondered?

"Holy Mother of God! But look at the height of that bloody rim everybody!" Mr. Howard then suddenly shouted.

"Well that does it, Dr. Grieg. We'll never be able

to climb that high without a rope."

"I know, Mr. Mallory," I said. "So I guess we have only two choices now: one is to walk around the circumference of this rim and try to find an opening. Or two, to hold that séance we were going to have."

"You left out a third choice, Dr. Grieg."

"And what's that Mr. Amundsen?" I asked.

"To question the other four people in here, myself included, because maybe we might learn something more - something the angel may want us to know before she frees us."

"Well I guess anything's better than fooling around with a séance if we can avoid it. So, Mr. Amundsen, what would you like to tell us of your experience with the angel you saw?" I asked him, as I noticed him begin to scratch his hawk-like nose and dry wrinkled face.

"I think I should first tell everybody about how I got lost at both Poles during my life as an explorer. It might just lead us to something important."

"Say anything you like, Mr. Amundsen," I answered him, praying at the same time he would have the key clue we all needed in getting us home.

"OK. Then let me start at the time of my second polar expedition. The year was then 1897, and I had signed on a ship named the *Belgica* as second mate. The expedition's leader was a Belgian by the name of Lieutenant Adrien de Gerlache. He wanted to land a small party on the Victoria Land Coast of Antarctica so he could make the claim of becoming the first to winter it out on that continent. But he also wanted to travel inland and survey the South Magnetic Pole. I was going to be in his select company.

"We had also aboard the ship the soon to be famous Dr. Frederick Cook, the ship's doctor. He had been with Robert Perry on his trip to Greenland five years before.

"Well, one night in January, while I had been on

watch duty at the ship's helm navigating through a narrow strait, I suddenly saw land appear through a heavy fog, so I immediately had to turn the ship in the opposite direction in order to avoid sinking the ship.

"Then later, after mapping the region where we landed, we continued sailing until we crossed the Antarctic Circle. However, in the month of February, de Gerlache, who was also a baron in his country, and a young one at that - a man of thirty-one, ordered the ship on into the farthest southern point the ship could sail. But the crew mistakenly believed we were instead heading back home, the second in command and I, along with de Gerlache, knowing quite the contrary.

"You see, as I mentioned, de Gerlache wanted to be the first explorer to winter it out in the Antarctic at any costs. Even so, by the beginning of March the ship was just over a hundred miles within the treacherous ice pack, with ice all around us and all channels to the open sea closed. We had hardly any food left on top of this, and no warm clothing except what we had on hand for the men of the shore party. And we didn't have enough lamps on the ship to light all the places we were going to need to light during the coming sunless winter months.

"However, I could have cared less, Dr. Grieg, since it was just this kind of difficultly that I craved. Because that's why I had signed up on the ship in the first place - to experience an adventure I would never forget.

"All the same, I never believed the ship would be frozen in the ice a total of thirteen freezing months, several of which the sun was no longer present; or that two of the men would go insane; or that scurvy would sweep the ship like a plague - one man, Dr. Cook, and I being only partly affected. Still, everyone survived except one man who had died from not wanting to eat the seal and penguin meat we had then obtained from our hunting. I was even in charge of the ship for a while when de Gerlache and the second in command became ill at this time. Hence, it was Dr. Cook who saved everyone, and within a week we were all almost as fit as before the outbreak of the scurvy - the vitamin C in the meat probably saving our lives.

"Even so, by the end of July my hair had turned gray, and I was only a man of twenty-six... Dr. Cook, in January of the next year, however, had tried to get the ship into a lead in the ice less than a mile away. He reasoned when the ice began to move, the channel would open up giving the ship an escape route to the open sea. Of course, we had to use four-foot saws and explosives to help this along. And after towing the ship through a canal to the lead, we waited for another whole month just within sight of the sea, before the lead widened and the ship was set free.

"All the same, just as the ship headed out toward the open water, it had to squeeze between two giant icebergs for several days, and Dr. Cook, now fearing the ship would break up against the sides of the icebergs, came up with the idea of using the skins of the penguins we had killed as makeshift mats, so as to cushion the sides of the ship from the impact of the ice. And by God! It worked like a charm. And that's about the whole story of my near fatal second expedition, Dr. Grieg."

"That's an unbelievable story Mr. Amundsen. But tell us now about your angel sighting, if you will?" I asked.

"This much I can tell you, Dr. Grieg: the year was 1903, and I was a man of thirty-one, and I had just spent my inheritance to buy a sixty-nine-foot sloop I renamed the *Gjoa*. I had purchased this craft in order to become the first man to successfully navigate the Northwest Passage.

"As you know, Dr. Grieg, the Northwest Passage is a shortcut through nine hundred miles of very tricky channels, between many islands in the Arctic from

Alaska to Greenland. There had already been over sixty failed attempts to find a route through this passage. Nevertheless, I knew I could find a way through where these others had failed, because everyone before me had used a large vessel, and with my small sloop I knew I would be able to push my way between the ice floes.

"Of course everyone thought I was mad and would never make it in such a small craft. Still, I made it through the Northwest Passage regardless. I had thus navigated 3450 miles through dangerous Arctic waters, a journey that had taken me almost three years to complete. Of course, I'm skipping over the many trials and tribulations my crew and I had to endure during these years."

"What of your angel sighting? You still haven't told us of it?" Mr. Howard said, as I noticed everyone nodding their heads in agreement.

"Yes, I was just about to tell you all. Well, what happened was this: after rediscovering the new location of the northern center of the earth's magnetic field, which had shifted, thus proving the theory of magnetic pole migration, I saw directly above this new spot, located somewhere at Cape Adelaide on Boothia Peninsula, what could only have been an angel from heaven materializing. How I knew what I was seeing was an angel, was for the fact that Peter Ristvedt, my companion, saw her as well."

"Then what happened?" I asked.

"We both saw her start to pray very close to us, after Ristvedt and I opened a little Bible and began reciting the Twenty-Third Psalm. We almost fainted when the angel began doing this because of the shock. Still, we were so happy we were alive, having succeeded in our mission that had taken us almost ten days, we forgot about the lightheaded sensation we were feeling. And thank God we didn't pass out in that subzero weather, because we would have surely died.

Anyway, after reciting the psalm, we saw this angel immediately float straight back up to a cloud, the only one in the sky, and disappear. We both, then and there, promised we would never speak of this experience to anyone. We knew people would thereafter voice doubts about our sanity. And that was the one thing most people already questioned in the first place... I mean what my companions and I were doing in this area of the world risking our lives. So our credibility had to be beyond reproach if we wanted people to believe we had actually discovered the North Pole's new location."

"Yes, Mr. Amundsen, we can agree with that. We all would have been called insane if we had related our angel sightings to the world. Now tell us how you ended up in here with us all? I believe, the last the world heard of you, you were in your plane looking for some Italian who had crashed in his Zeppelin?"

"Yes, Dr. Grieg, that's correct. His name was Umberto Nobile, the famous Italian dirigible designer. And the name of his supposedly crashed dirigible was the *Italia*. Unfortunately, my men and I never found him or the sight where he went down. And that was because we ourselves went down before finding him."

"Didn't you go down in a plane on another expedition with the American who put up the money?"

"Yes, Miss Earhart. His name was Lincoln Ellsworth. And he was in another plane that landed close to my own. We had landed a few miles apart believing we were at the North Pole. Unfortunately, it took us thirty days to get back into the air with the one plane we had finally freed from the ice. Still, just as we were flying within sight of the island of Spitsbergen, an aileron broke and we were forced to land the plane, which was a seaplane, on the choppy Arctic sea. Fortunately though, we managed to taxi the plane to shore, and within hours spotted a sail of a nearby sailing vessel, to which we had to again taxi to get their attention. When I got back home to Norway, I was welcomed home like a hero along with my crew. It was a reception like I never experienced before, not even after I had discovered the South Pole on my most famous expedition in 1911.

"But getting back to Mr. Nobile, if I may. Let me tell you of his first polar expedition I was unfortunately involved in, only three weeks after my return to Norway with Mr. Ellsworth. We were in a dirigible, this time named the *Norge*, which Mr. Ellsworth had purchased for one hundred thousand dollars, and where Nobile was aboard as the pilot, with one of my assistants, Riiser-Larsen, acting as his copilot.

"By the way, this was the expedition where the American explorer, Richard E. Byrd, along with his pilot, Floyd Bennett, challenged us to the North Pole, which Admiral Perry had earlier discovered. This is exactly what happened on my expedition to the South Pole, only in that one, the explorer, Robert Scott, had tried to beat me there. Unfortunately, as some of you know, it cost him and his men their lives.

"I'm telling you about this man, Nobile, because I want you to know why I decided to rescue him when he went down in his dirigible, the *Italia*, sometime later.

"As for this expedition in the *Norge*, I must say Nobile was an incredibly inept pilot. The most inept pilot I had ever seen. He didn't even know how to bring the damn dirigible he himself had designed in for a simple landing. I had only consented in letting him act as the pilot because the Norwegian Aero Club, whose president was overseeing the dirigible's refitting in Italy, had asked me to. Nevertheless, as I was to learn later, Nobile had met with him earlier and extracted concessions from him. Still, I wouldn't doubt if Mussolini, the Italian dictator at the time, had something to do with this, for I knew he probably wanted to use this historic flight to glorify his fascist country, as I figured Nobile wanted to glorify himself as well.

"As for the race to the North Pole with Admiral Byrd, I really wanted nothing to do with it. However, Nobile insisted upon the race. All the same, I held my ground and refused his requests. The result was that Byrd ended up being the first to fly over the North Pole, which I would have done with my American friend, Mr. Ellsworth, three weeks before, if we hadn't landed in the wrong spot by mistake. Still, I wonder if Byrd actually flew over the North Pole like he claimed because he may have only flown his plane to the point where Mr. Ellsworth and I mistook for the place.

"Anyway, I should have never let Nobile anywhere near the controls of the *Norge*, knowing he had designed a previous dirigible, the *Roma*, which had crashed three years before, killing twenty-four men in the process. Although I must say, he only commanded the *Norge's* steering, and was involved in the actual handling of the ship's elevator and vertical controls a mere two times. Yet he nearly cost us our lives.

"By the way, Mr. Ellsworth was with me on this expedition like before. And his job was to help out and measure the atmospheric electricity with a machine that was aboard.

"It was later, however, about three hours after we took off from King's Bay, Spitsbergen, that our airship slowly made a descent to about two hundred feet off the ice sheet. Still, we had to climb back up to about three thousand feet when ice began to form on the entire craft. Nevertheless, by the afternoon of the next day, we had made another descent not caring if we all froze, we were then directly over the North Pole, and I knew I had been the only man in history, along with Riiser-Larsen, to have reached both of the earth's poles successfully.

"Consequently, it was only two hours later that the dirigible ran into serious trouble. This happened when we crossed the hypothetical, or Inaccessible

Pole: the place where the Arctic ice is said to all drift. What happened was the entire area immediately became thick with fog, a fog so thick, it started to condense and form large chunks of ice all over the ship's skin. And then, what we all feared would happen abruptly occurred: these chunks began breaking off the skin of the ship, with some of them hitting the propellers of the engines, thereby causing the propellers to project these jagged chunks right into the ship's thin canvas. Moreover, we feared other jagged projectiles of ice would shortly pierce the hydrogen gas cells, thereby causing the ship to lose the very thing it needed to stay in the air.

"Then another thing happened even more serious: our radio reception began to grow fainter and fainter. We quickly realized this was because the ship's four-hundred-and-fifty-foot aerial was coated with a thick layer of ice. Even though we tried to clear this ice off the aerial many times, it finally broke off from the weight of the ice built up there. So we quickly replaced it with another aerial, but the same thing happened again, and we lost five more aerials within the next few hours: three from the ice, and two more when they were snapped off by the tips of two mountains.

"At about the same time, however, the tip of the dirigible began to nose downward because of the ice that had accumulated there, so we had to shift the gasoline ballast to level off this motion. We then discovered we were only twenty-one miles off our planned course of travel. About three hours later we finally spotted land off the port bow and knew we had crossed the Arctic.

"It was the next day, however, that we realized we were traveling over the Bering Strait, heading on a course directly to Siberia. So we had to turn the airship around and travel no more than six hundred feet above a turbulent sea, with winds so strong they were heaving the airship up and down a hundred and fifty feet at a time.

"It was within the next few hours, however, that Nobile took over the ship's controls in an effort to escape the fog we were flying in. This he managed to do by taking the airship up to an elevation of almost four thousand feet. It was at this time, unfortunately, that Riiser-Larsen left Nobile to return to his cabin. This was a big mistake because Nobile ended up not remembering to watch the airship's gas pressure, and because the airship was so high up now, the hydrogen began to rapidly expand due to the warming of the sun. This meant if we didn't take the airship down to a safer level immediately, we would burst its canvas envelop within a few minutes. However, we couldn't descend in enough time because the nose of the airship was then pointing up, so Nobile started to spin the elevator wheel to bring the nose down. But the wheel wouldn't respond, to his and everyone's surprise, so he began screaming for everyone to race to the airship's bow.

"Fortunately though, one of my men, Oscar Wisting, then turned the vertical helm control and the airship began to immediately tilt. This happened only after everyone, except two helmsmen and Nobile, had raced to the airship's nose carrying cases of food and cans of gasoline to make this area of the airship heavier. This resulted in quickly getting the nose down, thank the Lord. But then the same men had to race to the other end of the airship in order to prevent the airship from going into a nosedive. During all this time, the airship was being blown inland, and we had a hell of a time trying to get the airship on course fighting the terrible winds.

"On top of this, Nobile had the gall to then tell me he didn't consider any of my men sufficiently expert enough to pilot the airship any longer. So he took over the damn elevator control himself. And wouldn't you know it, Nobile again became distracted at the wheel, as the airship then changed direction due to his absentmindedness in changing the course. We would have crashed into a mountain if Riiser-Larsen hadn't acted fast by pushing Nobile aside and grabbing the wheel from him in time. We only cleared the surface of that mountain by a few lousy inches.

"It was soon after this that some of us began to see things that weren't there...."

"What exactly do you mean, Mr. Amundsen?" I asked him, sort of spellbound by his gripping tale, and now wondering what he was about to disclose.

"I don't know exactly, Dr. Grieg, because most of us were exhausted from all that had transpired. But some of us saw what looked like an army of men and horses directly below us. They appeared to be moving at a fast gallop, and there seemed to be a ghostly white haze all about them as well; we immediately realized what we were witnessing had to be a mirage. It was then brought to my attention by one of the helmsmen that in 1885, more than six hundred troops of the French colonial forces, who had been marching toward Saigon in Indo-China, mysteriously disappeared fifteen miles from that city. They weren't even under attack at the time, and not one of their bodies was ever found. So maybe what we had seen was only their ghosts down there on the ice..."

"You know, Mr. Amundsen," Mr. Howard was saying, "in 1939, the same thing happened to nearly three thousand Chinese troops stationed south of Nanking, China. All the search parties ever found when they looked for them were some of their guns, equipment, and cooking fires that had gone out. Of course, they could have all deserted before being attacked by the Japanese, who had just captured Nanking, but it's unlikely all three thousand of them would have done so.

"On the other hand, there's the case of the four thousand soldiers who disappeared in the mountains of the Pyrenees during the Spanish War of Succession.

No one ever found a single body or any equipment belonging to these men either. Something just swallowed them up I guess. Perhaps it was the cloud that brought those Nazis and British soldiers from Gallipoli to this place, the cloud from which our angel friend floated down and pointed. So maybe, Mr. Miller, you should blow your trombone again to get her attention once more. Because we seem to be just as lost as before she pointed this way."

"Mr. Howard," I said, "I think it would be better if we let Mr. Amundsen continue with his story, and then let Her Royal Majesty, Anastasia, and His Royal Highness, Tsar Alexander, tell us theirs. Because now that Dr. Rogers and Professor Moebius are here, what these last three people have to say may change our angel's mind about keeping us here. As to why she pointed this way, well it's obvious she wants us to know where we are, wherever that is. Or perhaps she just wanted to get us to an area in this volcano where we would be safe for now. So Mr. Amundsen, please continue with your story," I said to him.

"Of course, Dr. Grieg. Now where was I? Oh yes, I remember now...Well, after we had seen that mirage of soldiers below the airship, we picked up a radio bearing from the station in Nome, Alaska, learning in the process that the weather was calm and help was waiting for us there. It seemed the whole world believed we had perished because of our inoperable radio, for a search party was already being readied in Nome.

"Nevertheless, we were having a hell of a time trying to get there because of the weather. So we decided to land the ship in a village by the coast instead. If the wind had been blowing a little harder we would have all been killed in the process when an anchor we had lowered didn't take hold in the ice. This had occurred just as a sudden gust of wind hurled the ship toward the frozen shore of the village. All the same, I thank God the wind let up in time for the anchor to take hold down below.

"Anyway, after the villagers ran to grab the ends of the landing ropes we had cast over the side, and the crew shimmied down the rip chords and joined them, Nobile, having already stopped the airship's two engines and released the hydrogen in the gas cells, watched as the ship began to drift down to the ice like a paper plane. This had happened not more than a hundred yards from the nearest native dwelling, I might add.

"All in all, we had taken about seventy-one hours and traveled over three thousand and ninety miles after leaving Spitsbergen to reach this village, which was about a hundred miles from our destination of Nome.

"Even so, let me tell you what Nobile did after we landed in this village. He began sending telegrams to the newspapers claiming he had been a co-commander of the expedition, which was a lie. And altogether, he even went so far as to claim he had the whole responsibility of the voyage. You can't imagine what else he did to make me and my men look like fools. Why, he was even sent to America on a damn lecture tour by Mussolini, after Mussolini promoted him from a colonel to a damn general. Poor Mr. Ellsworth, my co-commander, the very man who had put up the money to buy the dirigible was made to appear by Nobile as nothing more than a mere passenger.

"So you can understand, Dr. Grieg, why I disliked this man, Nobile, from the first moment I had the misfortune of meeting him. When I learned two years later that he had set out on an expedition to the Arctic Ocean on his own, I almost laughed myself to death at the news.

"However, when I was informed he had crashed his dirigible on the second day of his voyage, I knew immediately I would be asked by my country to go and find him. What other choice did I really have in the matter, unless I wanted to be looked upon as a coward by everyone? But I had published a denunciation of this man in my memoirs the previous year, and Mussolini didn't approve of my help in finding Nobile. So my country of Norway, not wanting to offend 'Il Duce' told me not to go, asking my friend, Riiser-Larsen, to go instead. Nevertheless, I had to go find Nobile at any cost, for my honor was at stake and I had publicly announced that I would. Unfortunately, I had no money of my own to organize a rescue mission so the French government agreed to provide me with a seaplane after a fellow Norwegian businessman convinced them it would add to their country's prestige.

"The plane the French gave me was the one Rene Guilbaud, the French pilot, was going to use to attempt to cross the Atlantic. It was a Latham 47, and was so new; it had been little tested, mostly because it was a long-range type of aircraft.

"So on June 28, 1928, at the age of fifty-six, I took off in the plane with five in my crew from Tromso, Norway, to find Nobile. Unfortunately, within two hours in the air, after entering a fog bank, the plane developed a fuel leak somewhere in the line, and we had to land in the open sea. Then, just as I was going to use my radio to send out an SOS, the plane was blown over and started to sink. So the crew and I had no choice but to hang on to the floats of the plane's undercarriage, which stayed above the water's surface, I thank God. Even so, we knew unless help arrived within a couple of hours we would be all dead from exposure because of our completely frozen clothes.

"It was then, Dr. Grieg that a miracle occurred to our complete astonishment: a cloud began to descend directly above us, drying out our nearly frozen clothes in the process. This I believe also happened at the 'Miracle of Fatima', where over one hundred thousand people had their clothes dried out when the Virgin Mary appeared in the sky after it had rained...Anyway, the next thing I knew, I was in this place looking for my five-man crew, who had strangely disappeared from sight. And that, Dr. Grieg, is my whole story as I remember it. And I hope I didn't bore you and the others with the many details of it. Nevertheless, I had to set the record straight, in here especially, what with our angel friend being so near, not to mention, maybe the Lord."

"Yes, I know, Mr. Amundsen. However, let me also ask you if you can think of any connections you might have with anyone here, other than your angel sighting and the cloud, anything at all?"

"I was waiting for you to ask me that question, Dr. Grieg. And now that you have, I'm afraid I can't answer you, other than the fact I had to acquire the art of mountain climbing in my youth, which would connect me with Mr. Mallory. But Dr. Grieg, aren't you of Norwegian descent, like myself?"

"Why, yes, I am."

"Well then, there's another connection. And oh yes, I may have something in common with Miss Earhart as well."

"Oh? What's that, Mr. Amundsen?" everyone heard Miss Earhart ask.

"Simply that I was the first person in Norway to obtain a civilian pilot's license, as I already mentioned to Mr. Kennedy and you. I was a man of forty-two at the time."

"Wait a second, Mr. Amundsen," Miss Earhart was saying. "I just remembered, one of the men who was a member of your crew on the *Norge* was a technical advisor for my flight across the Atlantic in 1932 - four years after your disappearance in fact. His name was, Bernt Belchen."

"That's quite remarkable, Miss Earhart...you know, he was with Richard A. Byrd and Floyd Bennett on their expedition to the South Pole."

"Yes, I know. He was also with your American

friend, Lincoln Ellsworth, on a flight over the Arctic in 1931. Unfortunately, Balchen made an enemy of Mr. Ellsworth when Belchen refused to continue flying their plane when dangerous clouds appeared a mile or so ahead. And I believe Balchen's words on landing the plane were: 'Ellsworth can commit suicide if he likes, but he can't take me with him.' Still, they discovered five islands, five fjords, and several mountain peaks on their flight."

"Well, Miss Earhart, Mr. Ellsworth was always an adventurous type of man. Would you happen to know what became of my good friend after my disappearance and his flight?"

"No. I'm afraid not, Mr. Amundsen."

"I believe I know," Mr. Rockefeller said, then added, "He unfortunately fell down a cliff in the Grand Canyon in 1939, and suffered a permanent head injury, then died of a heart attack in 1951...Oh, and there's a permanent exhibit devoted to him at the American Museum of Natural History. It's partly there because he made many contributions to the museum, Mr. Amundsen."

"Well he deserves an exhibit Mr. Rockefeller because without his financial assistance I could not have made two of my polar flights. And his ranking as an explorer is right up there with the best of them, right along with me, I might add."

"Mr. Amundsen," I said, "if you can think of any other connections you might have with anyone here then feel free to interrupt me. Because I'd now like to ask Her Royal Majesty, the Grand Duchess Anastasia, to tell us her story next," I finished saying to him, as I then turned to Her Majesty and added, "If that's all right with you, Your Majesty?"

"Yes, Dr. Grieg I am ready to relate my encounter with the angel, which may take some time to fully say, as I suspect, His Royal Highness, Tsar Alexander's will as well."

"That's all right, Your Majesty. We seem to have plenty of time," I assured her, not really knowing how much time we really had before the volcano blew its lid, but knowing the angel had saved everyone's lives, I knew she would save us again when the volcano became life threatening to us.

"Hey! Wait a second," I then heard Elaine suddenly remark. "I remember seeing a play about a woman who claimed to be, Her Majesty, Anastasia. I even saw a film based on the play starring Ingrid Bergman, Helen Hayes, and Yul Brynner in 1956."

"Hey yeah! I saw that film also, Dr. Rogers!" Mr. Cooper next added.

"Me too! I saw the film as well."

"So did I, Mr. Hoffa," I said.

"But, Philip, don't you see what this means?"

"No? What, Elaine?" I said perplexed.

"It means that if that woman the play and film were based upon was in fact the real Anastasia, the woman who was claiming to be her years after her supposed death, then we, or at least you, Your Majesty, managed to get out of here and back to your time."

"Elaine," I said, "then why didn't Anastasia ever make mention of her experience in this place to anyone after escaping?"

"I believe it was because she partially lost her memory after a failed suicide attempt in Germany for some reason. I believe Your Majesty was in Germany looking for your mother's relatives, and it was because of failing to find them that you then decided to drown yourself by jumping off a bridge. However, this is what is so uncanny: I remember reading in a book about your life, Your Majesty, that one of your childhood playmates, a man by the name of Gleb Botkin, the son of your family's physician, a man you would later meet, and who would turn out to be one of your most loyal supporters in your struggle to prove who you were, was also the leader of a small sect devoted to the worship of the goddess Aphrodite!

"This book about you mentioned you never joined Mr. Botkin's church, instead, it related to the fact that all the letters he sent to you always ended with the prayer: 'May the Goddess bestow her tender caress on Your Imperial Highness's head.' Now isn't that an interesting and important, if not chilling, fact, Your Majesty?"

"Yes, it is. So I guess Gleb must have seen the angel as well. He probably thought she was the goddess Aphrodite, or he wouldn't have founded that church devoted to her...Do you remember, Dr. Rogers, how that book about me ended? I mean was I ever acknowledged as the daughter of Tsar Nicholas the Second? Or did I sadly die in obscurity, a penniless old and forgotten woman?"

"That, I'm afraid to say, Your Majesty. Still, whatever I tell you, you won't be able to remember on re-entering your time, as the book and film about you so evidently proved."

"Then don't tell me the answer to my question, Dr. Rogers. I just might refuse to leave this place when the time comes. And besides, from the look on your face, I already know what you would have replied..."

"Your Majesty, I think perhaps you should tell us of your angelic experience now. And anything else you might care to add, feel free to say," I said, as I noticed a look of wonder on Tsar Alexander's face.

"Well, Dr. Grieg, like I mentioned, there is a lot that's happened to me before I arrived here. And I must tell you and the others because I'm sure my great-great-granduncle, Tsar Alexander, is more than a little interested in hearing about my father and his sad and disastrous reign."

"Yes, I am, dear child," His Royal Highness said, as everyone watched him give Anastasia a loving smile.

"Yes, Your Majesty, tell us about your father, Tsar Nicholas," I asked her.

"First I must tell everyone that my father once owned a hollow gold cross filled with beeswax which had embedded in it a tiny fragment of wood from the very cross Jesus Christ was crucified on. The cross had been given papa at his baptism. Unfortunately, when papa was a boy he unknowingly swallowed the holy wooden fragment when he had eaten the beeswax after going to bed hungry one night.

"You see, when papa used to leave the dinner table, everyone had to follow him. My sisters and brother were always the last to be served when papa had guests, and we had to go without dinner sometimes like papa had as a child. I am telling you this about the cross because it may have something to do with why I am here with you now.

"Also, my name, 'Anastasia', means: 'The Resurrection'. - 'She will rise again'. And the fact that I have double lifelines in both of the palms of my hands, might be the reason mama decided on my name."

"That's very interesting, Your Majesty," Elaine was saying. "Still, didn't your father, the Tsar, once save your life from drowning when you were about five years old?"

"Yes, he did. So maybe that explains why I have double lifelines perhaps. Yet papa always wished he could have also saved the hundreds of people who had been accidentally killed at his coronation. What had happened, and this relates to papa's angel sighting, was this..."

"Wait a second, Your Majesty. You mean to tell us that your father also experienced seeing an angel?" I asked, somewhat bemused.

"Yes, he did, Dr. Grieg. This was at the coronation ball the very day of this terrible tragedy. The ball had been arranged by the French Ambassador, the Marquis de Montebello, and was to take place at the

Palace of Roses, which had been adorned with one hundred thousand roses sent by the French government for the occasion. However, papa did not want to attend the ball after all these people had died. They had been killed because of being trampled to death during a mad rush when someone among the crowd of more than five hundred thousand, yelled out there would only be enough free beer to those who got to the beer wagons first.

"Nevertheless, because of the urgings of his brothers and uncles, papa had given in to their pleas to attend the ball. Still, he had done so, so as not to offend the French who were Russia's only European ally at the time.

"Papa had been so distressed that night that he had decided to pay out of his own pocket, one thousand rubles to every family who had lost a loved one because of this tragic event. And it was when papa had gone off alone to his private study in the Rose Palace that he saw this angelic being appear. He told me about this experience years later. That is why I know what actually occurred. Papa told me he had only seen the angel for a few seconds, and that she had been crying. Mama herself began crying when papa retold this story to my sisters, brother and I. And it was then that I knew papa was telling us the truth about what he had seen..."

"Did your father ever have another such encounter with an angel, Your Majesty?"

"Why, yes, he did, Dr. Rogers, and I saw the angel as well. I was only a child of four at the time, and will never forget what I saw. I might add that I had yet another angelic encounter when I was fourteen, which was about three years ago. It occurred in my room at the Imperial Palace at Livadia, where there were many rose gardens laid out in large triangular beds. There was even a statue of the goddess Aphrodite on display in one of the gardens. And the entire palace was always filled with the sweet aroma of the roses during the summer months.

"However, let me tell you of papa's second angelic visitation. It happened at the *Palace of Roses* again, when I was playing near the palace's rose garden with my sisters and brother Alexis. It was a few minutes after papa learned of the sinking of eight battleships, seven cruisers, and nine destroyers at the *Battle of Tsushima* on May 27, 1905. This was during the Russo-Japanese War. There were over twenty eight thousand Russian soldiers and sailors killed in that horrible war, not to mention fifty-eight thousand Japanese...Anyway, papa and I saw the angel when I happened to follow him after he had received the sad news of this battle. It was just like that night after his coronation at the Rose Palace, papa wanting to be alone I mean.

"I never told papa I had followed him and witnessed this second angelic sighting of his. Nor for that matter, the second angel I observed as a teen...You see, I never understood what I had seen when I was four years old. It was only after my second sighting that I realized what I had beheld. However, there was a more important reason I never told papa of my angelic experiences. And that was because of that evil-minded man, the 'moujik' - Gregory Rasputin...

"You see, Dr. Grieg, I knew if I told papa, he would have told Rasputin. And Rasputin would have tried to make of my childhood encounters with the angel, the confirmation, or proof he needed that papa had not been hallucinating both of his sightings, for Rasputin had already such a strong hold over my mother, Alexandra, that I did not want to add anything to strengthen that hold. I say this because of what I learned about this so-called holy man..."

"And just what did you learn about Rasputin, Your Majesty?" I asked.

"Well first, I learned that the name 'Rasputin' in

Russian means - 'desolate', that his real name was - 'Efimovich'. And he was like his new name implied, truly desolate in his moral behavior, which he never dared exhibit around me or my family.

"Bishop Theophan, however, the saintly Inspector of the Theological Academy, the bishop who had first recommended this Rasputin to my mother, the Empress Alexandra, because of being supposedly impressed by Rasputin's faith, shortly thereafter began entertaining grave doubts about Rasputin's holiness. Bishop Theophan was then hearing the confessions of dozens of women who had gone to Rasputin for his spiritual help and had received instead, orders to undress and make love to him on the spot; Rasputin had knavishly told these complying women they had to commit a sin before they could be saved by the Lord.

"This 'goloni', or bearded bum, for that was what Rasputin really was, had, and I must confess, one redeemable quality as regards to my family, specifically my brother, Alexis. And that was that he could mysteriously stop the flow of Alexis' blood whenever Alexis injured himself. As most of you already know, my brother Alexis suffered from the dreaded blood disease of hemophilia.

"How Rasputin had been able to stop Alexis' blood from flowing, I later learned, was due not to the practice of hypnotism, which I might add, he only learned after his first meeting with Alexis, but rather to the calming effect he had upon Alexis' impressionable mind. Rasputin had accomplished this by telling Alexis stories in the dark with such a commanding voice he cast a kind of spell over Alexis, making him forget about the pain he had been suffering due to the swelling effects of his bleeding.

"This emotional change which Rasputin caused in my brother's mind and body evidently made my brother's blood slow down and eventually stop altogether within the area he had been bruised. Since

Rasputin was the only person able to prevent Alexis from bleeding to death, you can understand why my mother, Alexandra, listened to and unfortunately obeyed everything Rasputin had to say, almost as if he were the Lord God Himself.

"And it was because of Rasputin's incredible influence over my mother, that papa ended up having to abdicate as the Tsar of all the Russias, being later shot to death with all of my family..."

"Why was that? Just what did this Rasputin character do to cause your father's abdication and death?" we all heard Mr. Hoffa say as everyone watched Anastasia wipe the tears of sadness flowing down her face.

"He advised mama to appoint ministers, officials, generals, and anybody he liked to positions they were not qualified for. Thus he eventually brought about, through his meddling, the deaths of thousands of Russian soldiers during the World War, and what was to change my beloved country forever: *The Russian Revolution.*"

"What of your second angel sighting, Your Majesty? Can you please tell everyone about this now?" Mr. Howard was heard to ask.

"Well, it had to do with that devil Rasputin again. And it happened at the Imperial Palace at Livadia, near Yalta, like I mentioned. What happened was this: it was the day before Easter Sunday and I had just come into my room on the second floor after having gone swimming in the Black Sea - the very sea where I had almost drowned to death as a child, when suddenly Rasputin walked into my room unannounced. I was at that moment completely naked and the scoundrel seemed not in the least concerned by this embarrassing fact. As I began to cover myself, I saw him strangely begin to cross himself, for appearing directly behind me outside my window was the very same angel I had seen years before. When I turned around to look at her, I saw that she was pointing her left index finger straight at Rasputin's evil face. But when I turned back around to look at that pompous ass of a man, he had disappeared from my room, and so had the angel who had been hovering at my window - she was likewise nowhere to be seen.

"I later learned, at supper that night that at the precise moment of this angel's appearance to me, the portrait mama had of Marie Antoinette hanging in her room, had suddenly fallen to the floor. This mama had immediately taken to be a bad omen.

"Rasputin, on the other hand, well he never dared make mention of anything that had happened in my room to anyone, as far as I know, for if papa had ever learned Rasputin had come into my room when I had been completely undressed and had not left instantly, it would have been the firing squad for him. I, however, never told papa about this strange and frightful happening because of the embarrassment, as well as the concern I had for my brother Alexis. He needed Rasputin's strange ability to stop his bleeding when it became necessary, you see. Still, I wonder to this day what Rasputin would have done to me if it had not been for that guardian angel coming to my aid.

"As for what seeing this angel did to Rasputin's behavior from then on, well, he was never quite the same, he would always give me a look of outright fear whenever I appeared. Maybe he realized that by seeing this angel, an angel he most probably believed to be the 'angel of death', his allotted time on this earth was to shortly end. And it was not very long after seeing the angel, nine months to be exact, that he did depart this earth. Because on December 29, 1915, Rasputin, as you know, was murdered by Prince Felix Yussoupov, Grand Duke Dmitry Pavlovich, who was the son of papa's last surviving uncle - Grand Duke Paul, Vladimir Purishkevich, who was a member of the Duma, which was the Imperial Parliament, an officer by the name of Sukhotin, and an army doctor named Lazovert..."

"You mean one of my future descendants, this Grand Duke Dmitry, was one of the men who murdered this Rasputin, my child?" everyone heard His Royal Highness ask.

"Yes great-great-granduncle, he was unfortunately."

"It is of no consequence then, for I would have surely done likewise, my pretty one. And as you probably know from the books on my life, I also had such a person as this Rasputin who I listened to, much like your mother, dear child. However, in my case, this person was a very cunning woman - the Baroness Julie von Krudener, a prophet of sorts, whom I foolishly let influence my decisions."

"Oh yes, great-great-granduncle, I have read of this woman quite recently. But she was a saint compared to the likes of Rasputin, he must have been the devil himself to have not immediately died from the poison cake Prince Yussoupov had given to him, but Rasputin remained alive for an additional two and a half hours. However, it was not the poison cake that finally killed him, nor the bullet Prince Yussoupov fired into Rasputin's back, great-great-granduncle, rather it was these facts I am sure you would like me to tell, if I may?"

"Yes, my child, please do. Confession is good for the soul, be it yours or Prince Yussoupov's."

"Well, what happened from what I have been told, great-great-granduncle, was this: having been shot, Rasputin, who had been pronounced dead on the floor by Dr. Lazovert, who had checked Rasputin's pulse, suddenly and without warning, sprung to his feet and grabbed Prince Yussoupov by the throat with one hand while he proceeded to rip off one of the Prince's shoulder epaulets with the other hand.. Whereupon, the Prince, now fearing for his life, made it to the basement stairs leading to the first floor of the Moika Palace with Rasputin right behind him in a rage.

"Rasputin, however, had not actually been chasing the poor Prince, rather he had been heading for the iron gate of the palace across the snow-covered courtyard leading to the street. Grand Duke Dmitry, who was on the first floor, on hearing the Prince yelling that Rasputin was still alive and right behind him, then almost crashed into Purishkevich, who was at the stairs. Fearing for his life, the Grand Duke immediately fled into his parent's apartment to hide. Purishkevich, however, now realizing Rasputin was getting away, immediately made a mad dash after him firing two shots from his pistol, completely missing Rasputin's back. After firing a third time and hitting Rasputin in the shoulders, Purishkevich fired his pistol again hitting Rasputin in the head. On reaching him now on his knees in the snowy courtyard, Purishkevich, in a mad panic, began repeatedly kicking Rasputin in the temple of the head.

"Prince Yussoupov, who had now reached Purishkevich carrying a rubber club in his hands, then began to hit Rasputin again and again. Finally, having rolled up Rasputin's dead body in a blue curtain, which they afterwards bound up with rope, Prince Yussoupov and the others secretly carried Rasputin to the frozen Neva River, where they dumped him into a hole in the ice.

"Nevertheless, three days later, when Rasputin's lifeless corpse was discovered by divers searching for him in the Neva, it was learned he had actually died not from the bullets that had hit him, but from being drowned instead, because when these divers brought Rasputin up from his watery grave, they discovered he had freed one of his arms from inside the roped up curtain, as impossible as this must sound.

"So you can understand, great-great-granduncle, Miss Earhart and gentlemen, why I believe

Rasputin was not a normal human being, but rather a man possessed by the devil himself…"

"Yes, you're probably right about that, Your Majesty," I replied. "Still, tell us how you survived the firing squad that killed your entire family. Was it because of our angel perhaps, Your Majesty, the one you saw two times before?" I asked as she began to twist a handkerchief she was holding into knots.

"That I cannot be certain of, Dr. Grieg. Still, I would not be at all surprised if she was directly responsible for why I was spared. However, what really saved me was my poor dear sister, Tatiana, falling on top of me after she was shot. What then happened to me I really do not know, except for the fact I felt a terrible blow to my head a few seconds later, the result of a rifle butt hitting me.

"You see, I had fainted just as Tatiana had been shot. The men of the firing squad must have thought I had also been shot as I fell to the ground. Nevertheless, I was moments later bleeding to death as I lay unconscious, for when I awoke, after being rescued by two brothers who had been part of the firing squad, my right foot, left arm, and right lower chest were all bandaged.

"I later learned it had been these two brothers who had inflicted these wounds upon me. They had been ordered to finish off my family and I with their bayonets, but had taken pity on me when they had noticed I was still alive. So they immediately decided to stab me lightly in non-lethal areas so as to draw blood and mask the fact I had not been shot. And that is when I had to be hit in the head by a rifle butt by one of the brothers, I later learned. They did not want me to be conscious and scream while being stabbed, you see, I would have been instantly shot by the others in the firing squad, most definitely…"

"Why were you and your family being executed in the first place, Your Majesty?" everyone heard Miss

Earhart ask, her beautiful eyes looking at Anastasia like a concerned mother's.

"It was because of the 'White Army Loyalists' who were then advancing on Ekaterinburg - the place my family and I were being held captive by the Bolshevicks. The Red Army commanders feared we would be rescued, so they immediately decided to kill us rather than put papa on trial as they had been planning..."

Everyone then noticed Anastasia begin to cry, as I also observed His Royal Highness suddenly approach her and place his hands upon her shoulders.

"There, there, my little one," I heard him say, then add, "Don't cry. You're suppose to be a Romanov, and Romanovs who are to be the Tsars and Tsarinas do not cry, so you must be strong, as well as thankful you had such a loving mama and papa...My father, Tsar Paul, was a tyrant, who I even consented to have dethroned for the good of the empire, I have paid heavily ever since for this deed...My poor judgment back then resulted in his death, you see."

Having heard Tsar Alexander relate this revealing fact in his life, I took the opportunity of saying to him, "Your Royal Highness, if you would like to tell us your story now, including when and where you were during your encounter with the angel, we would be greatly honored. And please feel free to tell everyone about your meetings with the Emperor Napoleon, whom you defeated. I'm sure all of us would love to hear you speak about him."

"Yes, Dr. Grieg, I am sure all of you would. However, do we really have the time? Should we not now be trying to summon the angel again and get her to complete the job of leading us home?"

"We can do that a little later, Your Royal Highness. Still, I'm pretty sure she'll show herself again after we've all finished telling our stories. And like you told Her Majesty: 'confession is good for the soul', I'm also sure when the time comes for this volcano to blow its stack, our angel friend will come…Isn't that what's likely to happen, Professor Moebius?" I turned to him and said.

"Very probably, Philip, but you should have asked Dr. Rogers that question instead."

"Well, Elaine, what do you think will happen?"

"I think she'll show herself Philip, like you mentioned - with or without Mr. Miller's trombone solo, I might add. As to what she's likely to do when that occurs is anybody's guess. Still, there's a good possibility she will help us get back to our times or she wouldn't have pointed the way to this crater wall."

"Yes, Elaine, you must be right," I said, as I turned back to Tsar Alexander and said, "Your Royal Highness, please tell us of your life as it relates to what's happened to us all."

"Well, Dr. Grieg, ladies and gentlemen, I am dressed in the attire of a monk as you can see. Why this is so has to do with what happened to my father, as well as what later happened in my reign as the Tsar of all the Russias. However, I will first start by telling you of my grandmother, the Empress Catherine the Great, whose policies I detested as far as her ruling the empire was concerned. As for my father, Tsar Paul, he actually hated his mother, Catherine, with a vengeance, so much so, he always avoided being in her presence whenever possible. It was not unreasonable then, Catherine passed him over as her successor in her will in my favor instead…"

"Your Royal Highness, why did your father hate his mother so very much?" everyone heard Miss Earhart ask.

"It was because of her allowing the rumor to spread that my father was her illegitimate son. It may have been true for all I know; Catherine had many lovers during her reign. And also, she may have directed the conspiracy of having her husband, my grandfather, Grand Duke Peter, murdered. This my father believed because Catherine had not seen fit to punish the assassins thereafter.

"As to what happened when Catherine died, well I knew beforehand she had named me as her successor, however, because I did not want to defy my father, for I lacked the courage, I sadly declined the throne in his favor.

"Nevertheless, it was within days I knew I had made a terrible mistake..."

"Why was that, Your Royal Highness?" Mr. Wallenberg asked.

"Well first my father had his father's coffin disinterred from the vaults of the Alexander Nevsky Monastery, and then placed beside Catherine's, which was resting in state with her body in the Hall of Columns of the Winter Palace. This he had done because he wanted to give his father the royal funeral he had never received. And to add the proper touch to this funeral, my father even had the man chiefly responsible for his father's murder lead the procession from the Winter Palace to the Cathedral of Saint Peter and Saint Paul, one Alexis Orlov, as well as his accomplices, Passek and Barialinsky, as two of grandfather's pallbearers. This Alexis Orlov, father even made walk ahead of the coffin bareheaded in the subzero weather carrying the gilt cushion with grandfather's crown...

"I am telling you these facts to demonstrate the kind of man my father was, because if you are all to understand why he was eventually murdered, I must relate these and other facts to you. My reason for this is, if I am here in this place because of my part in my father's death, no matter how innocent I believe myself to be, then you must know, including God and this angel, why my father was killed."

"Yes, Your Royal Highness, tell us what in your heart you must," I said, "because if our angel is indeed hearing us speak, let alone God Himself, then perhaps they will realize you, as well as the rest of us, with the exception of Herr Bormann and Mr. Dillinger, are here by some mistake," I finished saying, knowing Bormann and Dillinger were now eying me with a maddening look.

"Yes, Dr. Grieg, it must be by mistake we are all here, for hell itself could not be more depressing than this timeless domain. Nevertheless, let me continue telling you why all of Russia considered my father a tyrant during his reign. You see, it was a few months after he had become the Tsar, and after he had done some good for the country, such as: founding a school of medicine, disallowing the serfs from working on Sundays and holidays, granting subsidies to new manufacturers that he began to be suspicious of everyone. By this I mean, anything that seemed to him subversive, such as: people being out of their customary dress, he could not tolerate in the slightest, so he ordered that no one would be allowed to wear top boots, laced shoes or straight pants. And to enforce this law of his, he even had his dragoons go out into the streets of Saint Petersburg ripping off waistcoats, tearing off hats, and collecting the shoes of the offenders.

"You see, Dr. Grieg, my father wanted everyone to wear three-cornered hats, knee breeches, buckles shoes, stand-up collars, and powdered wigs. He also wanted to regulate what everyone was reading as well, for he feared a revolution would result if people were allowed to read whatever they wished..."

"We Nazis burned all the subversive books in Germany as well, Your Royal Highness."

"Keep quiet, Bormann!" I shouted to him. "Your turn will come."

"Dr. Grieg, Herr Bormann has just proven that truth is always the first casualty of any war or revolution...Nevertheless, as I was saying about my father, he was obviously losing his mind at that time, he then allowed a newly appointed general of the Prebrazhensky regiment - the very regiment of my father's bodyguard, a man by the name of Alexis Arakeheyev, to bite the noses, tear off the mustaches, and slap the faces of any soldier under his command. This man, who I would later come to rely upon, was the most uncultured, ignorant, and mean commander this regiment ever had.

"However, it was at this time my father began to suspect me of plotting against him. And I must say he had good reason, for I was having secret meetings with several of my friends, one of which had just come back from the city of Paris. In our discussions, we talked of the benefits of the French Revolution, which had recently taken place, and how we could, when I became the Tsar, enact those same democratic policies which were making a better life for the people of France. The reason I was having these meetings was because Saint Petersburg had become, in the four years my father had been the Tsar, a vast prison surrounded by guard posts, with plainclothes police spying on everyone, and I and my friends wanted to stop this insanity from continuing.

"Nevertheless, unbeknownst to us all, there were other secret meetings going on in Saint Petersburg, by men who were plotting against my father with the intentions of removing him from his throne. They feared a war was very imminent with England over Malta, which England had failed to restore to the Maltese Knights, and which the English had promised to do. These other countrymen of mine knew the English had a much more superior navy to my father's. They also knew if a war broke out with the English, because of the English ships that would be seized inside of Russian waters, with their crews immediately imprisoned by order of my father, that Russia would have no choice but to quickly capitulate.

“However, what happened to my father shortly thereafter is I am sorry to say, almost the same as happened to this man - ‘Rasputin’, for one of the plotters in this group of more than fifty men, most of whom I knew, was one of father’s most trusted advisors - Count Peter Alexeyevich Pahlen. He had even informed me of his plans to remove my father as the Tsar, knowing I had been having secret meetings of my own discussing this very thing. I had at first refused any approval of his secret plot. Nevertheless, after I learned my father had no intentions of making me his heir, I reluctantly agreed to the Count’s plan.

“I concurred with him that no harm was to come to my father, and that on the signing of his abdication, he would be allowed to retire to one of his estates near Saint Petersburg, together with my mother, or even a mistress if he so desired. And if this were all done as he promised, I would not denounce him or his men in return.

“All the same, on the night of Count Pahlen’s coup d’état, after my father had finished supper and retired to his private apartment in the Mikhailovsky Castle, all fifty of Count Pahlen’s men, consisting of grenadiers, horse guards, naval officers, artillerymen, and men representing all of the regiments of the garrison, men who were fed up with the humiliation and shame my father had brought upon them and their homeland, did to my father what I so inwardly feared they would do. It was when entering my father’s apartment in the castle that ten of these men, who were all officers, suddenly burst into my father’s bedchamber unannounced. It was a few minutes later, however, after they had all failed to get my father to sign his abdication, that two of the men present, one being a general, then hurriedly left father’s apartment to get Count Pahlen, who was nowhere to be found. It was during this time that the remaining officers in father’s bedchamber suddenly lost their temper, when father began screaming before and after being hit in the temple of his head by a gold snuffbox someone in the room had angrily thrown. It was at this very moment father was sadly murdered, for now fearing my father's frantic screaming, one of the officers suddenly grabbed a sash off of another officer next to him and began to coil it around father's neck until it was tight. Need I mention when Count Pahlen finally entered father's bedchamber a few moments later, father was already dead at the Count's feet. Count Pahlen had obviously not wanted to be in father's apartment if father ended up being murdered due to a struggle with these officers. Still, if Count Pahlen had been there, my father, in all likelihood, would not have been killed.

"I was, during this time however, in my apartment on the ground floor waiting the coup's outcome. My mother, nonetheless, was trying to get into father's apartment to see if he was alive, all the time being held back by the officers. When word of my father's death reached me by way of Count Pahlen himself, I immediately broke down and sobbed. It was only because of my wife who was by my side, together with the Count, that I stopped crying like a child, for the Count had said to me in an unyielding voice to go show myself to his men and the guards, and then act as their rightful Tsar, which I of course immediately did…"

"Your Royal Highness," everyone now heard Mr. Cooper say, "What did you tell the people of Russia your father died of? I mean how were you going to protect these officers who were involved with his death?"

"I told my countrymen and the world my father had died following an attack of apoplexy. I also told everyone I would reign the same way as my grandmother, Catherine the Great, had done, and what a relief that was to all of my people and my wife."

"Your Royal Highness, was that about the time you saw the angel, or was it before that sad event in your life?"

"It was after my father's death, Monsieur Saint-Exupéry, and it was your country's most famous son who was responsible for my angelic encounter, I might add."

"You are of course referring to the Emperor Napoleon, Your Royal Highness."

"Yes, indeed I am, monsieur, and he also had such an encounter with something of a similar nature. Still, from what he told me on our first meeting during a ceasefire at Tilsit, after my army's defeat at the Battle of Friedland, where over twenty-six thousand of my soldiers had been killed, whatever Napoleon had seen had been anything but an angel, it was instead a being he referred to as: 'The Red Man'. This 'Red Man' had appeared to him inside of the Great Pyramid of Giza, after he had gone into the King's Chamber alone. However, Napoleon stubbornly refused to tell me what had transpired in there with him. As far as my angel sighting is concerned, well I saw her while I was in the Cathedral of Kazen, where I had gone by myself to pray. She appeared directly in front of the high holy alter just beneath the crucifix of Christ, with her left side to me and praying with her hands together..."

"Your Royal Highness," I said, "how could you be certain at that moment that this figure of a woman was in fact an angel you were seeing?"

"I knew she was an angel because I could see short rays of light coming from her head which gave her the appearance of wearing a crown, plus the fact she had suddenly appeared out of nowhere, as I was looking at the wooden figure of Christ. However, this is what convinced me on the spot of her divinity: she was knelling on what appeared to be a small white cloud! And in what could have been no more than a minute after her materialization, I saw her and this cloud then mysteriously float straight up and disappear, exactly as Mr. Kennedy said his angel had done in Saint Peter's

Basilica, of which the church of Our Lady of Kazen in Saint Petersburg is an exact replica."

"Great-great-granduncle, why had you gone into that church to pray?" Anastasia then asked.

"It was because of Napoleon, as I mentioned, my child, right after I learned of the burning of my beloved Moscow. I had entered this church in Saint Petersburg with the people not uttering a single cheer on my appearance. The morale had sunk to such a low level at this time that I knew the people were already aware Napoleon had taken Moscow.

"It was because of my angelic experience, which God had surely been responsible for, that I became a monk some fourteen years later. It was immediately after seeing this angel, however, when I had lost my honor and the country's, thereby bringing ruin upon its people, that I vowed I would not rest until either I or Napoleon were defeated or dead.

"At the same time, I thank the Lord God Almighty that the Russian winter did for my people what my army and I could never have done, for of the three hundred thousand men Napoleon had with him when entering my country, he had only thirty thousand men left when entering his own.

"It was three years later, notwithstanding, after I had read a German mystical book by the curious title of *A Cloud over the Sanctuary*', by Karl von Eckartshausen, that I realized I had not as yet found the inner peace I had been searching so long. This book, given to me by my friend, Koshelev, who was the Grand Master of my Imperial Court, and who was a Freemason and my spiritual advisor, was a book whose title could not but remind me of the cloud and angel I had seen in the Holy Sanctuary of Our Lady of Kazen. It was this book together with my angel sighting that was the turning point in my life, because it was then that I began preparing myself for the day I would disappear into the countryside where I would secretly devote my remaining years to the worshipful study of God. This wish of mine to disappear from the world was not only desired by myself, but by the people of my country in the last six years of my reign, I am sorry to say."

"Why was that, Your Royal Highness?" Mr. Mallory asked.

"First was because I had become almost like my father. You see, I had in 1807, reestablished his secret police. I had done this in order to keep a closer watch on the mood of my people, and because of the unrest going on in the country resulting from my delegating the internal affairs of Russia to of all people, General Arakcheyev - the general who had been in charge of my palace guards and had abused them. Nevertheless, he was the most loyal man I had at the time, and I trusted him like no other man in my domain. He knew how to get things done. He had a will of steel and a passion for work. True, he was a coward on the battlefield, uneducated and cruel, but everyone trembled in his presence. And though I knew he insulted, humiliated, and mocked most everyone he came in contact with, I nonetheless needed him as an umbrella, or buffer between me and my people. Still, by being my principal aid and administrator, he turned most of the provinces of the country into military colonies, strengthening the police surveillance and censorship simultaneously. This of course caused many rebellions to regrettably occur.

"However, what I did in 1822 caused the most unrest among my people; I reinstated the right of the lords to consign their serfs to Siberia, a right I had previously abolished in 1809."

"What of the reforms brought about by the French Revolution you promised you were going to try after becoming the Tsar, Your Royal Highness?"

"Well, Monsieur de Saint-Exupéry, your country's republican principles could not be applied to the brutal Russian reality, as I unfortunately learned, I am afraid, for as much as I wanted to apply those reforms, I knew they would never work, given the power of the landowners, the aristocracy and such.

"You see, Dr. Grieg, ladies and gentlemen, I was always terrified of change, of anything that would upset my reign because of the passing of liberal measures. This is why I founded the 'Holy Alliance', whose member nations, under the influence of Austria's Chancellor Metternich, advocated the concept of monarchy by divine right, for the aristocracy of the Alliance feared revolution among their people, who were demanding the right to self-determination during this time.

"You see, I had, when becoming the Tsar, the idealism of the Encyclopedists, Voltaire, and Rousseau, much like my grandmother, Empress Catherine, had before me. However, those beliefs I shared with her quickly began turning to a conviction of even stronger autocracy, as had hers after a brief period of her reign.

"Nevertheless, because of the repression of the serfs, the people of the cities, the men and officers of my army and navy, secret societies began springing up throughout the length of the land. The leaders of these societies, I learned, were mostly officers who had fought in the Napoleonic wars, having as a result, caught the fever of Napoleon's novel and liberal ideas. They naturally wanted to free Russia of all slavery in any form, as they had done for the rest of Europe's people because of the war. And after a rebellion of one of my palace regiments occurred, the ranks of one of these secret societies began swelling to an inexplicable amount. It was then I knew undoubtedly, the time had arrived to feign my own death and disappear as I had always planned, I knew if I did not do this I would be killed by my officers, as had my father during the four years of his sorry reign.

"How I then managed to feign my own death happened when a courier of mine named Maskov was accidentally killed while following behind my coach. He had died instantly when his two-wheeled carriage struck a protruding mound and overturned.

"The idea of trading places with my dead courier came to me the following day, since I was by then ill with fever and very weak, so weak in fact that I had many fainting spells whenever I got up from bed to walk. My doctor had been so alarmed by this he thought I had little time remaining to live, so taking the opportunity of his mistaken prognosis, I pretended I was getting even weaker, whereupon the fool doctor began applying dozens of bloodsucking leeches behind my neck and ears, and administering two enemas to me the next day.

"Then having had enough of my doctor and his antiquated cure-alls, I reluctantly took him into my confidence and told him of my plan of switching poor Maskov's body in place of my own, with orders my coffin was not to be opened by anyone until it was safely back in Saint Petersburg. I knew by then Maskov's body and face would be unrecognizable due to the long two-month trip from Taganrog, the many stopping off points for holy masses that had been planned being the reason for this period of time.

"As to who was to be my heir, I had a daughter who died at the age of three, so I had to choose my brother, Nicholas, for my other brother, Constantine, had earlier asked not to be my successor..."

"What of your wife, Your Royal highness? How was she going to live alone? Did she not object to you being gone?"

"Well, Miss Earhart, my wife, Elizabeth, also feigned her death six months later so she could join me in my secret retreat. She was only forty-seven at the time and I forty-eight. However, as you can see, I am an old man of seventy now."

"How did you get here, Your Royal Highness? Was it a cloud you found yourself in on your arrival in this barbecue pit?"

"Yes, it was, Mr. Hoffa. I had been out walking near my secluded house in the outskirts of Krasnoufimsk, in the province of Perm, when all of a sudden a cloud came down from the sky and completely enveloped me. And like everyone else here, I found myself in this place without a clue as to why. But maybe I should tell everyone what I had been looking at before this cloud suddenly appeared, Dr. Grieg?"

"Yes, please do, Your Royal Highness," I said with great interest.

"As hard as this is to believe, Dr. Grieg, I had been looking at what seemed to be a man dressed in some sort of military armor of the kind worn by the Spanish conquistadors, which consisted of a plumed helmet and breastplate, I believe. He was also dressed in a long cape and had on a sword. But I only saw him for a few seconds before he dematerialized from sight."

"Hey! Wait a second! I've heard of a story very similar to the one His Royal Highness just related Dr. Grieg."

"Oh really Judge Crater? Please tell everyone about it then," we all heard the professor remark.

"Yes, of course. Well I think that man His Royal Highness saw was a soldier stationed in the city of Manila in the Philippines in 1593, believe it or not. He was supposed to have suddenly vanished from that city, only to appear in Mexico City twenty hours later. And on arriving there, he then reported for guard duty in front of that city's palace. He was then questioned because of his strange manner of dress by the authorities. He then told them he had been instructed that very morning to report for duty at the palace in Manila, saying also that the governor of the Philippines had been killed the night before. Well you better believe everyone who heard his story thought he was completely out of his mind. In any event, two months later, after he had related his story and been thrown into jail, news reached Mexico City that totally confirmed his fantastic tale. So the authorities had no other choice but to release him and put him on the next ship for the Philippines, so he could tell the same crazy story to his superiors.

"It seems, Your Royal Highness that you may have seen this very man. Be that as it may, how he could have appeared to you over two hundred and fifty years into the future, for the few brief seconds you saw him, is about as baffling as how he got to Mexico City and how we ended up in here."

"Maybe that's the very key we've been looking for, Philip," Elaine turned to me and said.

"What do you mean?" I asked, wondering what she was thinking.

"Well maybe we're here because of some error after all. Maybe our angel got her orders mixed-up from the 'Mother Archetype'," Elaine replied, as she then looked up at the cloud above, as if suddenly becoming aware of some cosmic mystery she had just solved.

"You mean from God, Elaine?" I said, now thinking of Claude Rains' film portrayal of Mr. Jordan and his mix-up with the actor Robert Montgomery.

"Well who else, Philip? The god 'ZEUS'! The father of Aphrodite!" Elaine almost shouted.

"Then we've got to tell our angel some way. But perhaps she already knows her mistake with regards to us, with the possible exception of Herr Bormann."

"And me, Dr. Grieg?"

"Perhaps Mr. Dillinger. However, you acted in self-defense when you killed that cop who was shooting at you. Still, murder is murder in the eyes of the Lord, so I'm afraid..."

"Herr Grieg," Bormann interrupted me, "so what are you planning to do now? Tell Herr Miller to blow his horn perhaps?"

"Yes, if his lips are up to it, but not before I hear you tell us your version of how you got here. And I'm warning you, if you don't tell us your story now, I'm not going to let you come with us should we find our way home. You'll just have to rot in this place forever."

"All right, Herr Doctor, I will talk. All the same, do I really have a choice in the matter?"

"No, not really, Herr Bormann, you don't," I said with disgust.

"Then I'll begin by telling everyone...Wait a minute, what about you Herr Grieg? You haven't told anyone about your experience. Don't you think you should tell us how you got here with that time machine of yours?"

"After you've spoken, Herr Bormann, only after you've spoken," I told him with a sneer.

"All right then, I'll hold you to your promise. Still, you might be interested to know, Herr Grieg, that as early as World War One the German Army used to use artificial smoke on which to project an image of the Virgin Mary with her arms outstretched in a gesture of peace. Our military did this over the trenches with a powerful beam of light in an attempt to discombobulate and put the British on the run. This we accomplished with great effectiveness at the famous 'Battle of Mons'. So perhaps your angel is nothing more than this. I would not be at all surprised if Reichsmarschall Hermann Goering was responsible for that angel we just saw back there."

"I doubt that very much, Herr Bormann, unless Herr Goering was alive in 1980 and could project her image inside of a nautilus shell, among the other places our friends have said. So quit stalling and tell me how you arrived in here, or else," I sternly told him, as I drew my right index finger across my neck for effect.

"Ja voll, mein fuhrer!" Bormann said as he gave me the Nazi salute, clicked his boot heels and added,

"Your wish is my command. Still, I do not really have much to add to what I have heard everyone say. You see, my plane on which my bodyguard of SS and I were on was hit by machinegun fire while flying over the English Channel toward neutral Sweden, so my men and I had to bail out...In any event, as we were parachuting toward the water below, we all entered a bright, glowing cloud. And in what was only seconds, I found myself alone in here and separated from my men, whom you already know ended up being gunned down by those British soldiers. And as far as this angel you all have seen is concerned, well I never had the privilege of seeing her until I arrived in here with you. But I did see this 'Red Man' whom His Royal Highness said Napoleon saw while he had been in the Great Pyramid, if that means anything to you, Herr Grieg?"

"It just might, Herr Bormann, so why don't you tell all of us this experience of yours."

"Do we have the time, as some of us have said, because if this volcano on top of which we are standing is in the process of coming to life, then what use is there in telling you anything. We will probably all die in half an hour anyway from the heat and molten lava when it melts this silvery floor."

"I doubt that Herr Bormann, so tell us anyway about this 'Red Man' you saw. I can always tell Mr. Miller to blow his trombone when the time comes, like I said."

"As you wish, Herr Grieg. I saw this 'Red Man' in the bunker in the Berlin Chancellery along with Herr Hitler himself during the last hour he was alive. That is why Hitler had decided to kill himself and not come along with me when I escaped from there. Of course, I did not want that senile old fool coming with me anyway. He would have only made my escape more difficult, and with him dead I would become the new Fuhrer of Nazi Germany, what was left of it that is.

"Still, I was planning on reforming the party in

South America, so I knew I had to escape Germany without Hitler hindering my flight. And since I was Hitler's personal secretary and administered his personal finances, I was in the position of transferring all of the party's plundered gold deposits there without needing his help.

"As for what this 'Red Man' did when Hitler and I saw him materialize, well this being just stood before us not saying so much as a word. It was as though he had been enjoying our mortal fear of him as Hitler and I were speechless and trembling in our boots. Who this 'Red Man' was, Hitler had no idea. I learned later, however, after my escape that Hitler had seen this 'Red Man' before, and that the 'Red Man' might have been an Atlantian from the lost continent of Atlantis. I knew this because the Atlantians were said to have been a red-skinned people, the same as your American Indian people are. He must have been a scientist from the past traveling in some kind of time machine or something, much the same as you have accomplished, Herr Grieg. Then again, maybe this 'Red Man' was the devil himself."

Before I could answer Herr Bormann, I suddenly heard the yelling of many individual voices in the distant fog. I was about to comment on this, being perhaps the other people in the crater yet unknown to us getting roasted alive, when I saw a black man dressed in only a kind of jock strap and holding a spear in his hand suddenly come out from the fog a hundred feet away.

"Hey, look everybody, I see a man carrying a spear over there!" I shouted, pointing his way.

"My God!" Mr. Rockefeller screamed. "That's one of the headhunters who were chasing me in New Guinea, Dr. Grieg! He must be searching for me now!"

"Then everyone with a gun get ready to shoot if he comes any closer," I shouted.

Almost at once, everyone saw at least fifty more

of these headhunters come out from the fog about a hundred feet away.

"Hey, Rockefeller!" I heard Mr. Hoffa shout. "This 'payback' thing you were talking about, well what if we give those cannibals Herr Bormann? Wouldn't that satisfy their 'payback' custom of revenge?"

"Yes, it probably would, Mr. Hoffa, but are we that sadistic and unciviilized? What would our angel friend say?"

"Yes, I suppose you're right, Doc, but if it should come down to it, I'm sure as hell going to sacrifice that Nazi pig to save our loving skins."

It was seconds after Mr. Hoffa had spoken, that all of us suddenly saw the silver floor of the crater where these headhunters were standing give way.

"Look!" Elaine shouted to me. "Those cannibals, they've disappeared! The floor of the crater must have melted right out from under them!"

"Yes, you're right!" I said loudly, as I frantically turned to Mr. Miller and said, "I think the moment has come to blow that trombone of yours again, my friend."

"That's for sure, Doc. That's for sure," Mr. Miller answered, as I watched him put his trombone to his lips just as a spear shot out of nowhere and bounced off the silvery floor striking him with its flat side squarely on his mouth. As I watched in horror his blood start flowing down from his split and bleeding lips, I shouted for everyone to hear, "Quick! Someone help Mr. Miller! He's been hurt!"

Immediately I saw Miss Earhart remove from her leather jacket a red scarf, which she then gently placed across Mr. Miller's bleeding mouth.

"I was once a nurse in a Canadian military hospital while a pre-med student at college, Mr. Miller," we all heard her say.

Mr. Cooper then turned to me and said, "I guess we're going to have to hold that séance, eh Doc?"

"Yes, I guess so, Mr. Cooper," I replied, as I

walked over to Mr. Saint-Exupéry and said, "Mr. Cooper is right. Now I'm going to have to hold a séance to summon our angel here. You seem to have had some experience along with Miss Earhart in this regard, so you're going to have to instruct me now."

"Don't worry, Dr. Grieg," he replied, "there is nothing to it. All we have to do is to form a circle with all our hands joined. Then all of us who have seen the angel before arriving here have to pray for her to appear, so that would exclude Herr Bormann and Mr. Dillinger."

"Doesn't someone have to act as the 'medium', as they say?"

"No, not if we are going to pray. In any event, you had better hurry with this séance because in about twenty minutes or less, this area we're standing on will be a boiling hot soup of silver."

Wasting no time, I did as Mr. Saint-Exupéry said and gathered everyone around, with the exception of Bormann and Dillinger, into one big circle where we then joined hands.

"OK, Doc, now what?"

"We pray for our angel to appear, Mr. Hoffa," I said. "And I think if we recite the twenty-third psalm of David now, as Mr. Amundsen did while at the North Pole, our angel will materialize as well."

"Yes, Dr. Grieg, she most likely will. I should have thought of that now."

"All right Mr. Amundsen, since you know this particular psalm by heart, why don't you say it out loud so those of us who haven't learned it can follow you."

"I think it would be a better idea if His Royal Highness did the honor, because if I'm not mistaken, I believe it was His Royal Highness who founded the Russian Bible Society."

"Is that true, Your Royal Highness?" I asked.

"Yes, this is true, Dr. Grieg, along with my friend Koshelev."

"Then please recite this psalm for us. We'll all repeat each of its verses after you," I said, as His Royal Highness began saying:

"The Lord is my shepherd; I shall not want," we all repeated except Bormann. "He maketh me to lie down in green pastures: he leadeth me beside the still waters," we repeated. "He restoreth my soul: he leadeth me in the paths of righteosness for his name's sake," we repeated. "Yea, though I walk through the valley of the shadow of death, I will fear no evil: for thou art with me; thy rod and thy staff comfort me," we repeated. "Thou preparest a table before me in the presence of mine enemies: thou anointest my head with oil; my cup runneth over," we repeated. "Surely goodness and mercy shall follow me all the days of my life: and I will dwell in the house of the Lord forever. Amen," we all repeated.

Then almost instantly, we saw, not the angel appear as we were so hoping, but two men coming out of the fog near the rim of the crater a hundred feet away. They were both dressed in aviator garb: leather bomber jackets, boots and helmets, glass goggles and khaki-colored slacks. As they began waving to us and frantically shouting something that was most definitely in French, Mr. Saint-Exupéry began yelling hysterically, "Mon Dieu! Mon Dieu!"

Yelling back to him I said, "Who are they? Do you know them?"

"Oui oui, Dr. Grieg! I know them both! They were given up for dead when their planes disappeared. They are my best friends. The taller man is Jean Mermoz, and the shorter man is Henri Guillaument....Jean! Henri!" everyone heard Mr. Saint-Exupéry start shouting to them again. "It's Antoine! Antoine de Saint-Exupéry!"

However, just as Mr. Saint-Exupéry's two friends arrived within a few feet of him, all of us saw about two dozen other men come out of the fog directly behind them.

"My God! Dr. Grieg!" Miss Earhart shouted to me. "There's my copilot, Freddie Noonan!"

"And there's the pilot and colonel who were on my plane!" Mr. Miller next exclaimed.

Followed by Mr. Kennedy rejoicing, "And there's my copilot, Wilford Willy!"

"Good God! There's the man I was with on Mount Everest, Mr. Andrew S. Irvine!" Mr. Mallory next frantically added.

This followed by Anastasia happily shouting, "There are the two guards who saved me from the firing squad, the Tschaikovsky brothers!"

"And look there, Dr. Grieg!" Mr. Amundsen was yelling to me next. "There are the men of my crew I had given up for dead."

And as all these men came up behind Mr. Saint-Exupéry's two lost friends, who were now embracing him and kissing him on the cheeks, I saw, as did the people in my group, three men whom nobody had as yet recognized suddenly step toward me and Professor Moebius as if in a daze.

"And what's your name?" I asked the first man to my left.

"My name is Colonel Percy Fawcett, the archeologist. And not more than a few minutes ago my men and I were exploring in the Mato Grasso of Brazil. Can you gentlemen please tell me where we are?"

"A bit later perhaps, Colonel Fawcett, I'm a mite pressed for time, I'm afraid. And your name, sir?" I said to the man in the middle.

"The name's Ambrose Bierce, the writer, and a few minutes ago I was in Mexico fighting alongside my pal, Pancho Villa."

"And you sir? What might your name be?" I finally asked the third man to my right.

"I'm Macmillan, of the famous publishing house. And I was just standing atop Mount Olympus a moment ago, believe it or not. What the hell's going on here? Who are you people?"

"Well I guess everyone is alive here who was dropped in this crater with the possible exception of those British soldiers who saved us from those krauts, eh, Dr. Grieg?" Mr. Hoffa was saying. "So I guess all we need now is a damn flying saucer to come rescue us, since our little angel friend is nowhere to be seen."

"Don't look now, Mr. Hoffa," Mr. Amundsen was saying in a worried voice, "but that cloud is now coming down from the sky."

As everyone looked up to the slowly descending cloud, it suddenly began thundering again and giving off bolts of lightening, as though a trainload of fireworks were now exploding in its milky white core.

Suddenly, the sound of a lone trumpet began to blast from above, and for a split second I thought Mr. Glenn Miller had just blown his trombone despite his cut and bleeding lips.

It was then Elaine turned to me in terror and said, "Philip, didn't thunder and lightning and the sound of a trumpet occur just as the prophet Moses was receiving the 'Ten Commandments' from God?"

"Yes, Elaine, it did, but then that would mean God is…"

"A woman and about to appear, dear Philip?"

Before I could answer Elaine, the cloud was already down upon everyone's heads, and I knew by the way I was feeling an instant later that the majestic glory of the Almighty Lord was indeed in our presence, for I was now feeling like a charged-up battery again, exactly as I had been when I had first entered this world.

However, instead of undergoing the unbearable pain I had previously experienced, I was now in the opposite state: my whole body and mind feeling like a torrent of orgasms were occurring one after the other, and I was wondering if any of the people around me were feeling in any way the same.

"Elaine!" I shouted to her, her face and body now hidden from me in the white misty cloud.

"I'm right beside you, Philip! Reach for my hand!" I heard her murmur; her voice now strangely calm and subdued, as I frantically took her moist hand into my own, immediately feeling as though I were in the air with her by my side.

"Elaine!" I shouted to her again. "We're floating off the ground! Don't let go of my hand! We must be leaving the crater now! God must be taking us home! Pray it's back to our time!"

Within seconds, I noticed that my feet were then strangely coming down on what appeared to be a bed of white roses, as the cloud Elaine and I were in instantly dissipated into a fine and vapory mist. It was then I realized with astonishment just where we had landed.

"My God!" Elaine yelled happily. "We're in the rose garden of the White House! And so is everyone from the crater. But who's that man in the white toga standing shoeless with... Is that the President of the United States with him?"

"I wouldn't be at all surprised, Elaine," I said, "or a bit surprised if that man in the white toga is the Sicilian philosopher, Empedocles, who disappeared atop Mount Etna ... From the dates on all the license plates I see, it must be the year 2012 ... Where's Mr. Mallory, Tsar Alexander and Anastasia? I wonder."

"They must have been left behind, Philip. Still, everyone else seems to be here."

"Yes," I said, "even that Nazi bastard, Herr Bormann, dead over there with a spear in his chest. Looks like those cannibals got their 'payback' it would seem."

Epilogue

And now, having set the record straight for you regarding my ordeal and adventure with the angel, and being as knowledgeable as myself about the famous people I have met, I shall conclude my narrative by telling you I married my beautiful Elaine a week after our return. The famous Mr. Amundsen, together with Amelia Earhart, are at the moment training as astronauts for the next mission to the moon. Mr. Antoine de Saint-Exupéry is now back in France writing another book on what happened to him and giving thought of perhaps running for political office, as is the young Mr. Kennedy in his native State of Massachusetts, and the even younger Mr. Rockefeller in his State of New York. Mr. D.B. Cooper and Mr. Dillinger are in jail rehearsing a revival of the play - *The Petrified Forest*, with Mr. Leslie Howard kindly acting as the guest director. Judge Crater has been appointed a justice of the United States Supreme Court and Mr. Glenn Miller and his newly formed orchestra have just been hired to take the place of the band on NBC's Tonight Show. Mr. Raoul Wallenberg is now campaigning for president in his country of Sweden after having just retuned from being honored in the country of Israel, and Mr. Hoffa has just been appointed the new Secretary of Transportation by the President of the United States.

As far as His Royal Highness Tsar Alexander the First and the dear Anastasia and her two lifesaving friends are concerned, well it seems they were all sent back to where they had been before arriving in the crater. Why this is so, I do not know, but it is my guess the woman claiming to have been Anastasia in that film about her was an impostor, as I believe it has already been proved, since the real Anastasia's bones and her brother, Alexis', were found in a shallow pit in 1991 with a perfect 'DNA' match when compared with the bones of the rest of her family which were buried in a separate shallow pit and also discovered in 1991. Plus the fact history would have been changed to a degree that might have resulted in a different Russia had the real Anastasia not been sent back to her time. Mr. Mallory, unfortunately, was also returned to his time and it is my guess it was because of his great unfulfilled desire to be the first to conquer Mount Everest, for he was recently found dead on Mount Everest very close to its top. However, because a photograph of his wife, Ruth, which he had promised to leave at the summit of Mount Everest was missing from his wallet, it is now believed Mr. Mallory was the first man to have climbed that awesome mountain after all. He must have fallen to his death on his way down, the world can now only assume.

As to what will happen to me, now that I have killed Professor Moebius, who still rests at my feet, well...wait just a second, I think I can hear him breathing now.

"Professor Moebius!" I shouted. "I'm so sorry you were knocked unconscious when I pushed you to the floor, but I couldn't allow you to stop me from blowing up the shell," I now said to him, thankful he was alive.

"That's all right, Philip. I know you didn't mean to hurt me," he said to me kindly, as I immediately began helping him to his feet with a great heartfelt joy.

"Professor Moebius, I thought I had killed you. Can you ever forgive me for having pushed you so hard?" I said, in a voice that was close to sobbing, my eyes close to tears.

"Yes, Philip, as I would have forgiven my own son if he'd asked. Still, you were completely right. I should never have tried to stop you from blowing up your shell. God only knows what might have happened if it had fallen into the wrong hands...Holy Mother of God, Philip! But your shell is still outside the lab! I thought you'd destroyed it?" the professor was shouting, now noticeably shaken and surprised, as I saw him look with startled eyes at my still existing marvel that had remained outside my lab the thirty-two years we had been gone.

"Yes, Professor, I was just preparing to blow up the shell again. But now that you're alive, why don't you do me the honor," I answered.

"Yes, dear boy, I believe I will," he replied, as I watched him walk to the switch that would set off the twenty sticks of dynamite I had placed in the shell.

As we both watched my 'Eighth Wonder of the World' shatter into a million pieces, we suddenly heard Elaine scream as she came running into the lab, "Philip! There's a giant mollusk in its nautilus shell coming out of the ocean! And its got legs!"

After which I calmly turned to Professor Moebius and said with a grin, "Why is it, Professor, there's never an angel around when you need one?"

I then heard him reverently reply,

"Or the 'Virgin Mary,' my dear boy?"

THE END

Acknowledgements

From: "Ces maisons qui tuent," Roger de Lafforest, Editions Robert Laffont 1970 and "L'art et la science de la chance," Roger de Lafforest, Editions Robert Laffont, 1968, I am grateful for permission to use excerpts from the above two books from the publisher: Editions Robert Laffont, Paris France, as well as the author, Roger de Lafforest in (Chapter Three) of my book, as well as the English translation of these two books above; "HOUSES THAT KILL," and "FATE CONTROL," translation by Lowell Bair, from FATE CONTROL by Roger de Lafforest, translated by Lowell Bair, copyright © 1974 by Berkley Publishing Corporation. Original © 1968 by Roger de Lafforest. Used by permission of Berkley Publishing Group, a division of Penguin Group (USA) Inc. Translation 2 quotes ca. 170 words, pp. 84-5, 86 EDITION: PBK Territory: World.

English translation by Lowell Bair, from HOUSES THAT KILL by Roger de Lafforest, translated by Lowell Bair, copyright © 1974 by Berkley Publishing Corporation. Original © 1970 by Roger de Lafforest. Used by permission of Berkley Publishing Group, a division of Penguin Group (USA) Inc. English translation; 7 quotes ca. 385, pp. 34, 54, 115, 126-7, 154, 154-5, 155 EDITION: PBK. Territory: World.

I would like to also express many thanks to Ms. Hilary Quinn, a great graphic web designer, for permission to use the photo of the Virgin Mary that adorns my novel's book cover. http://www.draiochtwebdesign.com

From "The Tao of Physics," by Fritjof Capra, copyright 1985 used by arrangement with Shambhala Publications Ltd. Boston, MA www.shambhala.com

I gratefully acknowledge the above publisher and author for permission to use these excerpts from "The Tao of Physics:"

"where one's individuality dissolves into an undifferentiated oneness."

"All things are aggregations of atoms that dance and by their movements produce sounds. When the rhythm of the dance changes, the sound it produces also changes. Each atom perpetually sings its song, and the sound at every moment creates dense and subtle forms."

"In the case of sound, for example, the air in a cavity will in general respond only weakly to a sound wave coming from outside, but will begin to "resonate," or vibrate very strongly, when the sound wave reaches a certain frequency called the resonance frequency."

"out of mind spring innumerable things, conditioned by discrimination. These things people accept as an external world. What appears to be external does not exist in reality; it is indeed mind that is seen as multiplicity; the body, property, and above-all these, I say are nothing but mind."

"Moebius strip riddle solved at last"
You can read the full article on this discovery reported by ABC News (Australian Broadcasting Corporation) on July 16, 2007 at:
http://www.abc.net.au/news/stories/2007/07/16/1979151.htm